BIOTECHNOLOGY IN THE MARINE SCIENCES

ADVANCES IN MARINE SCIENCE AND BIOTECHNOLOGY

Edited by
RITA R. COLWELL, University of Maryland
E.R. PARISER, Massachusetts Institute of Technology
ANTHONY J. SINSKEY, Massachusetts Institute of Technology

BIOTECHNOLOGY IN THE MARINE SCIENCES

Proceedings of the First Annual MIT Sea Grant Lecture and Seminar

Edited by

RITA R. COLWELL
University of Maryland

ANTHONY J. SINSKEY
Massachusetts Institute of Technology

E. RAY PARISER
Massachusetts Institute of Technology

A WILEY-INTERSCIENCE PUBLICATION
JOHN WILEY & SONS
New York ● Chichester ● Brisbane ● Toronto ● Singapore

Library of Congress Cataloging in Publication Data:

Main entry under title:
   Biotechnology in the marine sciences.

   Held Mar. 18–20, 1982.
   "A Wiley-Interscience publication."
   Includes indexes.
      1. Biotechnology—Congresses.   2. Aquaculture—Congresses.   3. Marine fouling organisms—Congresses.
4. Marine pharmacology—Congresses.   I. Colwell, Rita R.,
1934–   .   II. Pariser, Ernst R.   III. Sinskey,
Anthony J.   IV. Massachusetts Institute of Technology.
Sea Grant College Program.

TP248.2.B56   1984   600'.6'09162   84-2382
ISBN 0-471-88276-3

Printed in the United States of America

10 9 8 7 6 5 4 3 2 1

Contributors

John Abelson
Agouron Institute
La Jolla, California

Robert Belas
Agouron Institute
La Jolla, California

Gary J. Calton, President
Purification Engineering, Inc.
Columbia, Maryland

Donald R. Cheney, Professor
Department of Biology
Northeastern University
Boston, Massachusetts

Dan Cohn
Agouron Institute
La Jolla, California

Rita R. Colwell, Director
Sea Grant Program
University of Maryland
College Park, Maryland

Janet Elliott
30 Cobblestone Place
Wilton, Connecticut

William S. Gaither, Dean
College of Marine Studies
Sea Grant Director
University of Delaware
Newark, Delaware

Marcia Hilmen
Agouron Institute
La Jolla, California

Michael F. Holick, Professor
Department of Nutrition and Food
  Science
MIT
Cambridge, Massachusetts

Scott T. Kellogg
SIBIA
San Diego, California

Alexander M. Klibanov, Professor
Department of Nutrition and Food
  Science
MIT
Cambridge, Massachusetts

Peter J. Kretschmer
Celanese Research Corporation
Summit, New Jersey

Contributors

Robert Langer, Professor
Nutrition and Food Science
  Department
MIT
Cambridge, Massachusetts

Anne Lee
c/o Robert Langer
Nutrition and Food Science
  Department
MIT
Cambridge, Massachusetts

Alan Mileham
Agouron Institute
La Jolla, California

Richard Ogden
Agouron Institute
La Jolla, California

Bruce M. Poole, President
Sea Plantations, Inc.
Salem, Massachusetts

Donald W. Renn, Senior Research
  Fellow
FMC Corporation
Rockland, Maine

ChoKyun Rha, Professor
Department of Nutrition and Food
  Science
MIT
Cambridge, Massachusetts

Christophe Riboud, Professor
Institut de Gestion Internationale
Agro-Alimentaire
Cergy, France

Denise S. Richardson, Group
  Leader
Mogul Corporation
Chagrin Falls, Ohio

John H. Ryther
Harbor Branch Institute
Ft. Pierce, Florida

Michael Silverman
Agouron Institute
La Jolla, California

Melvin I. Simon
Biology Department
CALTEC
Pasadena, California

Scott Sindelar, Acquisition
  Analyst
De Kalb Agricultural Research
  Inc.
DeKalb, Illinois

John N. Vournakis, Professor
Department of Biology
University of Syracuse
Syracuse, New York

Harold H. Webber, President
Groton Bio-Industries
  Development Co.
Groton, Massachusetts

Robert L. Wetegrove, Group
  Leader
Water and Waste Treatment
Nalco Technical Center
Naperville, Illinois

George Whitesides, Professor
Department of Chemistry
Harvard University
Cambridge, Massachusetts

Preface

In 1972 the Massachusetts Institute of Technology Sea Grant College Program initiated an annual lecture to stimulate public discussion of critical marine issues of the day. In 1982, on the tenth anniversary of the lecture series, MIT Sea Grant expanded on its concept by adding two days of seminars to complement the lecture, at which experts from industry and academia could discuss the latest in research and business opportunities.

The March 1982 lecture and seminar, Biotechnology in the Marine Sciences, responded to increasing interest in the application of biotechnology to solve marine problems and create new opportunities for using marine resources. Over 200 people attended the lecture given by Rita R. Colwell, a noted marine microbiologist from the University of Maryland. About 155 industry and academic leaders participated in the accompanying two days of seminars, which were divided into four topics: biotechnology in aquaculture, marine pharmaceuticals and bioproducts, marine biofouling, and marine pollution control. At each seminar an overview of the technical and economic constraints and opportunities in the field was given, followed by papers and panel discussions on new research developments. The lecture and selected seminar papers are published in these proceedings.

RITA R. COLWELL
E. RAY PARISER
ANTHONY J. SINSKEY

College Park, Maryland
Cambridge, Massachusetts
Cambridge, Massachusetts
April 1984

Acknowledgements

We wish to take this opportunity to thank the National Sea Grant Program, MIT, and the John J. Wilson Endowment for providing funds for the Annual Sea Grant Lecture/Seminar Series. We would also like to acknowledge the important support we have received from the National Science Foundation, Messrs. Pennie and Edmonds, New York, and F. Eberstadt and Company, Inc., New York. In addition, a grant-in-aid from Verbatim Productions for recording and transcribing the proceedings is greatly appreciated.

The initiation, planning, organization, and management of a meeting such as this conference always requires a most taxing and protracted team effort involving many participants whose efforts and contributions it is impossible to acknowledge adequately. The names of at least a few persons should, however, be mentioned, since they carried the heaviest burdens on the long road from inception to completion of this project; they are Elizabeth Harding, manager of the communications and editing component of the MIT Sea Grant Program, together with her colleagues, Lynne Newman Lawson, Pat Gallen, Therese Moyer, Jacqueline Drapeau, Madeleine Hall-Arbor, and Molly Crowley. Without their untiring, careful, and professional assistance, this volume would not have seen the light of day.

R. R. C.
E. R. P.
A. J. S.

Introduction

Biotechnology and genetic engineering promise to provide novel solutions to biological problems that have challenged scientists for decades. Academic and industry researchers have seized this new opportunity, but until recently have tackled mainly land-related problems. This book contains papers from a conference on Biotechnology in the Marine Sciences (held by the Massachusetts Institute of Technology Sea Grant Program in March 1982) and represents a first attempt to look closely at a new synthesis of disciplines that is expected to lead to a more efficient use of ocean resources.

A paradigm has been formulated which defines biotechnology as a systems approach concerned with converting raw materials via biological transformation into useful products. The integration of genetics/biology, chemistry/biochemistry, and engineering/material sciences could increase and diversify marine food products, produce new marine-based pharmaceuticals, protect marine vessels from deterioration, and clean up environmental pollution. This book and the conference on which it is based have sought to illuminate and stimulate discussion of questions such as: What is the intellectual capital that supports biotechnology in the marine sciences? What are the products and processes upon which the researchers should focus? What is the most effective infrastructure through which scientists, engineers, and managers should work to further the development of marine-related biotechnological principles?

Part of the discussion focuses upon two main sources of intellectual capital that support the new field—rDNA and hybridoma technology. To date, these have made a number of developments possible:

1. Establishment of new and more powerful genetic systems for a wide variety of prokaryotic and eukaryotic organisms.
2. Development of new techniques for the directed manipulation of the genetic systems of microorganisms, plants, and animals generating new types of organisms with specific metabolic properties.
3. Evolution of new knowledge on the properties of enzymes that catalyze reactions of potential interest. It may soon be possible, using recombinant

DNA procedures, to generate new enzymes which catalyze unique chemical transformations to produce new molecules and materials.

4. Development of an aggressive high technology industry concerned with the use of modern biological methods for the production of chemicals. Fermentative production processes of the future will use newly designed microorganisms, cells of higher plants, and cells of human origin that can utilize a variety of substrates.

General Applications of Biotechnology to the Marine Sciences

How will these developments influence biotechnology specifically in the marine sciences? Both rDNA and hybridoma technologies are discussed in the book in general as well as in specific presentations forecasting this influence. R. Colwell gives a broad overview of both technologies as they relate to the marine sciences. More specific topics are covered by J. Vournakis, who discusses the role of rDNA technology of salt tolerance and heavy metal toxicity in plants. P. Kretschmer and S. Kellogg extend the principles of rDNA technology to the development of new genetic systems for algae and for the microbial control of pollution. D. Cheney presents an overview of new strategies that can be used to develop new and superior polysaccharide-producing algal strains. M. Simon clearly shows how modern developments in genetics can be used to further understand the scientific basis of marine biofouling processes. G. Calton discusses the use of monoclonal antibodies for the isolation of labile proteins of marine origin.

Specific Applications of Biotechnology to the Marine Sciences

One of the most difficult tasks currently facing scientists, engineers, and managers is how to use the new and powerful scientific tools appropriately to solve relevant problems. One of the pitfalls in the application of biotechnology concepts and techniques is, indeed, that they offer elegant and expensive solutions to trivial problems. Therefore, four specific and important areas in which the application of biotechnology promises to make a real difference were chosen for discussion: aquaculture, marine pharmaceuticals and bioproducts, marine biofouling, and marine pollution control.

Aquaculture. An overview of the economics of aquaculture, from the international viewpoint, is given by C. Riboud, while the more specific topic of aquaculture is presented by W. Gaither. S. Sindelar and H. Webber discuss the business potential of aquaculture. J. Ryther summarizes the application of biotechnology to mariculture problems.

Marine Pharmaceutical Bioproducts.   The paper by G. M. Whitesides and J. Elliott clearly outlines both short- and long-term research and development opportunities for developing organic chemicals from marine sources. Specific topics include chitosan as a biomaterial discussed by C. Rha. Chitosan, a unique biopolymer, can now be used in a variety of novel applications since many of its solution properties have been systematically determined. Biotechnological applications of algal products such as carrageenan and algin are summarized by D. Renn. The isolation and properties of provitamin D_2 and previtamin D_2 from *Emiliania huxleyi* and *Skeletonema menzelii* are discussed by M. Holick, while A. Lee and R. Langer present evidence that shark cartilage contains an inhibitor of tumor neovascularization.

Marine Biofouling.   In her presentation on the microbial ecology of biofouling, R. Colwell addresses many important aspects of this process that can now be explored by genetic methods. As noted above, M. Simon et al. show how modern genetic tools can be used to address the basic biology of the biotechnology process. D. Richardson presents a concise analysis of biofouling in freshwater cooling systems.

Marine Pollution Control.   The last topic covered is the role of biotechnology in marine pollution control. S. Kellogg describes how new genetic techniques, such as plasmid-assisted molecular breeding and state-of-the-art genetic techniques, are and can be applied to the development of new degradative bacteria combatting marine and terrestrial pollution. A novel application of horseradish peroxidase for the removal of hazardous organics from industrial aqueous effluents is presented by A. Klibanov; R. L. Wetegrove summarizes the microbial problems, solutions, and trends in industrial waste treatment, while B. M. Poole discusses the treatment technologies for the removal of biochemical oxygen demand.

   Many of the topics covered in this book will be dealt with in more detail in future conferences, workshops, and planning sessions to further define the applications of biotechnology in the marine sciences. Strategies, however, for building appropriate infrastructures must be designed by both the government and private sectors to capitalize more energetically on the opportunities in this field.

ANTHONY J. SINSKEY

Cambridge, Massachusetts
April 1984

Contents

LECTURE

Biotechnology in the Marine Sciences

Rita R. Colwell

University of Maryland
Sea Grant Program

Introduction

Dr. Kenneth A. Smith, Associate Provost and Vice President for
Research, Massachusetts Institute of Technology

   Welcome to the Massachusetts Institute of Technology, to the
Tenth Annual MIT Sea Grant Lecture, and to the first of a new
series of Sea Grant Seminars. Our annual Sea Grant Lecture was
initiated on the basis of our conviction that the awareness of
the public to significant marine-related developments and
problems must be increased, and to provide a forum for discussing perspectives and further uses of the seas. To achieve
this end, we have selected lecture topics that are wide in scope
and consonant with the interests of a general audience. I
believe that we have been successful in attaining those goals.
   Another of the continuing goals of the MIT Sea Grant Program
is to increase the effectiveness of our marine-related
scientific and engineering research programs. To do this, we
believe it is imperative that we contact and collaborate with
the research community and marine experts from industry, government, and academia. Therefore, we have decided to combine the
presentation of the annual Sea Grant Lecture with an in-depth
discussion of carefully selected marine research and development
issues, with the discussions to be led by selected experts who
are working at the cutting edge of their fields. Thus, today we
are initiating a new seminar series, a series which is the first
of its kind and, we believe, a model for the future.

For this occasion, we have chosen the topic of
"Biotechnology in the Marine Sciences." In part, the selection
is appropriate because of our many colleagues at MIT who are
actively engaged in this general area, and because progress in
the field of genetic engineering has attracted a great deal of
public attention and interest.

More importantly, biotechnology in the marine sciences has
not yet received the attention we feel it deserves. We are
convinced that the application of genetic engineering technologies to the manipulation of the marine environment can open new
fields and opportunities to both researchers and users. Such
exciting areas are, for example:

* Production of new pharmaceuticals, especially protein
  products such as vaccines
* Production of a host of organic compounds that appear to
  be available in large concentrations only in the marine
  environment
* Development of new, salt-tolerant food and feed crops

To initiate this first combined lecture and seminar series, and
to keynote the presentations and discussions that will follow
over the next two days, we are most fortunate to have Dr. Rita
Colwell of the University of Maryland as our tenth Sea Grant
lecturer.

Lecture

Modern science is taking a technological "leap" forward, an
advance so profound that it may alter the very structure of our
society. As occurred with the invention of the steam engine,
the discovery of the atom, and the advent of the computer--each
of which resulted in a revolutionizing of science and society--
the explosive development of biotechnology and its handmaiden,
"genetic engineering", will pervade all our lives.

Almost without being noticed, agriculture has incorporated
biotechnology into crop production, with plant-tissue culture
being employed for commercial production of ornamental plants
and for crop improvement, especially in the case of nitrogen
fixation with attempts to manipulate the Nif gene from bacteria
into plants. In the case of plant breeding, genetic engi-
neering--defined as the application of fundamental knowledge of
the molecular basis of biological processes to practical
industrial and medical problems--involves genetic transformations through cell fusion and insertion or modification of
genetic information via the cloning of DNA and its vectors. In
fact, techniques are now available for manipulating organs,
tissues, cells, or protoplasts in culture; for regenerating
plants; and for testing the genetic basis of new or unusual
traits. These methods, applied to plant breeding, have achieved
good success in growing cells in tissue culture into mature
plants, examples of which include the asparagus, citrus fruits,
pineapples, strawberries, and soybeans.

Headlines have announced the birth of "test-tube" babies

during the past few years. These babies were born after
externally fertilized eggs were implanted in the uterus. The
process represents an extraordinary advance, one made possible
by years of research in reproductive biology. Though none have
caught the public eye quite as has the human test-tube baby,
applications of breakthrough reproductive biology have also
yielded many significant improvements in domestic animal
production.

The use of artificial insemination--employing stored frozen
sperm, notably for beef cattle and probably for swine, sheep,
goats, and other animals--has perhaps proved to be the most
important recent technological development in reproductive
physiology. But until specific genes of farm animals can be
identified and located, direct gene manipulation may not be
practicable. Furthermore, most significant traits in farm
animals are due to multiple genes, creating a difficult, though
not insurmountable, problem.

Genetic engineering applied to the production of fish,
molluscs, and crustaceans in natural environments and hatchery
systems, although at the rudimentary stage, offers unique
promise. Applied to these animals, in vitro manipulations such
as cloning cell fusion, production of chimeras, and other
recombinant DNA techniques will provide an impetus for major
advances in applied genetics. In fact, Streisinger et al.
(1981) have already produced clones of homozygous diploid zebra
fish (Brachydanio rerio). Since successful aquaculture of many
species of invertebrate animals has been achieved, since large
populations of shellfish at the larval and intermediate stages
can be manipulated and their genes cloned, the stage is set for
the realization of genetic engineering's staggering potential.

The financial possibilities of genetic engineering have not
been lost on the stock market. Investors have embraced (nearly
smothered, in fact) new young companies with risk capital. Like
mushrooms, these companies have sprung up in large numbers and
in unexpected places. Without a doubt, gene manipulation can
open new opportunities for the marine sciences. Gene manipulation creates new combinations of heritable material by inserting
nucleic acid molecules (the basic genetic material of nearly all
organisms) into any virus, bacterial plasmid, or other vector
system--a transformation that incorporates into a host organism
heritable material it does not naturally possess but can now
recreate through natural propagation. Colloquially referred to
as "genetic engineering", these methods can be readily applied
to aquaculture of oysters and other molluscan species.

Sources of material lie readily at hand, thanks to
groundwork laid by the National Sea Grant College Program in
aquaculture. Existing facilities can handle projects aimed at
selection of disease-resistant stocks and at gene selection and
transposition. The technology is available. The challenge is
to bring together technology and opportunity.

The history of genetic engineering is fascinating, a story

told in many variations but in some scientific detail by a
number of authors, notably Professors R.W. Old and S.B. Primrose
of the University of Warwick, U.K., in a well-written primer,
Principles of Gene Manipulation: An Introduction to Genetic
Engineering, and in a bit more Chaucerian style by Professor
J.D. Watson in The Double Helix.

A story yet to be told resides in the extraordinary promise
and challenge genetic engineering holds for the marine
sciences. On the one hand, the potential of the world oceans to
feed and sustain humankind has been addressed in poetic terms
since the dawn of time, with impressive promises made during the
past few decades, including reports of a huge food source
represented by krill in Antarctic waters and by fishery stocks
in offshore waters of the world's continents. Indeed, the
legendary promise of the oceans to feed all of humankind may yet
be fulfilled, but if it is, it will most likely be via the
test-tube, by means of marine biotechnology.

Perhaps one of the most dramatic examples of biotechnological application is that of marine pharmaceuticals. In a 1977
conference on "Drugs and Food from the Sea: Myth or Reality,"
researchers described cardiotonic polypeptides from sea
anemones, an adrenergic compound from the sponge, Verongia
fistularis, and potential anticancer agents from Caribbean
gorgonians and soft corals. More recently, Rinehart et al.
(1981) have described antiviral and antitumor depsipeptides from
a Caribbean tunicate. Extracts prepared from the Caribbean
tunicate, an ascidian or sea squirt of the family Didemnidae,
inhibit growth of DNA and RNA viruses, as well as L1210 leukemic
cells. These depsipeptides—termed didemnins after the name of
the tunicate family, Didemnidae, from which they are isolated—
are closely related, but vary in activity. The discovery
indicates that the subphylum Tunicata or Urochordata (phylum
Chordata) may be an abundant source of bioactive compounds of
pharmaceutical interest (Rinehart et al. 1981). The tunicate of
the Trididemnum genus, when extracted with methanol-toluene
(3:1), showed activity against herpes simplex virus, type 1,
grown in CV-1 cells (monkey kidney tissue), indicating that the
extract inhibited the growth of the virus. This antiviral
activity may also involve antitumor activity. When tested
against other viruses, essentially all extracts of the tunicate
collected at a number of sites showed activity in inhibiting
both RNA and DNA viruses. The suggestion that the extracts
might also have antitumor properties was evidenced from their
high potency against L1210 murine leukemic cells. The novelty
of the didemnins results from a new structural unit for
depsipeptides, hydroxyisovalerylpropionate, and a new
stereoisomer of the highly unusual amino acid statine (Rinehart
et al. 1981).

The literature describes a variety of compounds from the sea
which act on the cardiovascular and central nervous systems.

This literature was recently reviewed by Kaul (1981), who has pointed out that drugs of high pharmacologic activity from nature have, in fact, been unsurpassed by synthetic compounds. Drugs from nature, predominantly from plants, include morphine, atropine, and digitalis glycosides, to name but a few. Marine animals and plants have yielded cardiovascular-active substances, and these include histamine and N-methylated histamines of sponges, viz. Verongia fistularis (Hollenbeak et al. 1976); asystolic nucleosides from the sponge, Dasychalina cyathina; and the nucleoside, spongosine, isolated from Cryptotethya crypta.

Several marine organisms have provided useful drugs: liver oil from some fish provides excellent sources of vitamins A and D; insulin has been extracted from whales and tuna fish; and the red alga, Digenia simplex, has long been used as an anthelminthic. Bacteriologists, for many years, have incorporated agar and alginic acids into laboratory media. In general, it has been uneconomical to extract and purify a drug from an organism which has to be captured in large quantities from remote corners of the world. Thus, only a few marine organisms are currently sources of useful drugs. Genetic engineering can change this situation dramatically, by opening up a vast and diverse range of marine life to probing for valuable pharmacological compounds. In the long run, these opportunities will develop as the tools for gene cloning are sharpened and the applications broadened.

Marine Toxins

Of particular interest are toxins produced by marine organisms. A toxin is a substance possessing a specific functional group arranged in the molecule(s) and showing strong physiological activity (Hashimoto 1979). A toxin has the potential to be applied as a drug or pharmacological reagent. Furthermore, even if direct use as a drug is not feasible because of potent or harmful side effects, the toxin can serve as a model for synthesis or improvement of other drugs. Many attempts have been made to develop useful drugs from the sea by screening for anticarcinogenic, antibiotic, growth-promoting (or inhibiting), hemolytic, analgetic, antispasmodic, hypotensive, and hypertensive agents.

Two successes demonstrate the potential. Tetrodotoxin, the main action of which is paralysis of peripheral nerves, is a valuable pharmacological reagent. Because it inhibits specifically the sodium permeability of nerve membranes, it has been valuable for elucidating the excitation mechanism. It must be emphasized that, for the moment, the applications of marine toxins are limited, to say the least, and it is mainly in the area of understanding functions that the toxins have an interest.

A second success is an insecticide developed from nereistoxin and widely marketed since 1966. Fishermen are

familiar with the fact that flies die when they come into
contact with the dead marine annelid, Lumbrinereis (Lumbri-
conereis) brevicirra, commonly used as bait. The toxin was
first isolated in 1934, and once its structure was determined, a
new insecticide was developed from nereistoxin. Cartap
hydrochloride is one of the synthesized derivatives. Active
against the rice stem borer and other insect pests, it does not
appear to be toxic to warm blooded animals, and resistant
strains of insects do not readily develop (Hashimoto 1979).

Toxins from marine animals that are well known have been
summarized recently and are listed in Table I. Hashimoto
(1979), in a recently published book, has compiled the extensive
literature summarizing information about marine organisms which
cause food poisoning or possess the capacity to produce a toxic
sting or bite or are poisonous, as in the case of the toxic
marine flagellates such as Gonyaulax and Gymnodinium spp. The
vast array of marine animals and plants covered in this
compendium includes algae, coelenterates (i.e., jellyfish),
echinoderms, corals, sea anemones, molluscs, algae, flagellates,
nemertines, annelids and other invertebrates, and fish.

Research on marine toxins or bioactive substances in marine
organisms has increased in recent years, with a number of
monographs and reviews appearing in the literature, including
Halstead, 1965, 1967, 1970; Russell and Saunders (eds.), 1967;
Baslow, 1969; Martin and Padilla (eds.), 1973; Scheuer, 1973;
Marderosian, 1969; Youngken and Shimizu, 1975; and Ruggieri,
1976. Work in this area has been the focus of symposia on
"Drugs from the Sea," held every two or three years since 1967.

Marine toxins show great promise not only as pharmacological
reagents, but also as models for the development of new
synthetic chemicals. Recently ciguatoxin, palytoxin, and
halitoxin have also been investigated and provide interesting
new information.

Ciguatera is a human disease caused by the ingestion of a
wide variety of coral reef fishes that contain toxins accumu-
lated via the marine food web. The principal toxin of ciguatera
poisoning is a heat-stable, lipid-soluble compound named
ciguatoxin by Scheuer et al. (1967), and the source of ciguatera
toxin(s) is a photosynthetic, benthic dinoflagellate,
Gambierdiscus toxicus Adachi and Fukuyo. The term ciguatera was
derived from a name used in the 18th century in the Spanish
Antilles for intoxication caused by ingestion of the "Cigua" or
turban shell Cittarium (Livona or Turbo) pica (Gudger 1930;
Lawrence et al. 1980). Occurrence of ciguatera was recorded as
early as the 1400s in the West Indies (Gudger 1930). The
subject of ciguatera food poisoning has been reviewed recently
by Withers (1982). Origin of the toxin is not yet fully
understood but may be derived by the fish from ingestion of
toxic tropical red tide dinoflagellates such as Pyrodinium
bahamense (Maclean 1979).

The major obstacle to solving the chemical structure of ciguatoxin is the problem of obtaining enough purified material for spectroscopic studies, since large quantities of toxic live eel yield approximately 1 mg of purified ciguatoxin from 1000 grams of starting material. The LD_{50} of ciguatoxin has been reported to be 0.45 ug/kg by Tachibana, similar to the range of palytoxin, which has an average LD_{50} of 25 ng/kg (nannograms per kilogram of bodyweight) in the rabbit, the most susceptible animal tested, to 40 ug/kg in the mouse (Moore et al. 1982). The primary action of ciguatoxin appears to be an increase in permeability of the excitable membranes to sodium, causing depolarization.

Because the source of ciguatoxin has not been clearly identified, application of gene cloning techniques is not yet feasible even if this compound shows potential for pharmaceutical use. Hybridoma technology, however, will prove invaluable for identifying, isolating, and characterizing this and other toxins, as well as for producing diagnostic and therapeutic reagents for treatment.

Hybridoma technology arose during the course of very basic, rather esoteric studies on the somatic cell genetics of immunoglobulin production by mouse myeloma cells. During experiments carried out to examine the regulation of expression of immunoglobulin genes, Kohler and Milstein (1975) fused cultured mouse myeloma cells to spleen cells of mice immunized with sheep red blood cells. Some of the hybrid cells containing the genetic material from both the myeloma and the normal spleen cells were found to produce antibody that reacted with sheep red blood cells. The antibody-forming cells not only grew continuously in culture but also maintained other characteristics of the myeloma parent, in that they could be frozen and recovered from the freezer. When injected into recipient animals, these cells formed tumors that secreted large amounts of the antibody molecule accumulating in the serum and ascites of the tumor-bearing animals. What Kohler and Milstein did was to immortalize a single antibody-forming cell, and the progeny of the cell continued to produce large amounts of the antibody molecule that it had been producing prior to the fusion (Scharff and Roberts 1981).

The main features of the technology are shown in Figure 1. Very briefly, spleen cells from an immunized animal and tissue-culture-adapted mouse myeloma cells are mixed together to associate randomly. Polyethylene glycol, which fuses cell membranes, is added to the mixture. Approximately 1 of every 2×10^5 spleen cells forms a viable hybrid with a mouse myeloma cell. Since most of the cells are spleen-to-spleen cell hybrids, myeloma-to-myeloma hybrids or unfused cells, the mixture is treated with hypoxanthine, aminopterin, and thiamine (HAT) (Littlefield 1964) to destroy the irrelevant cells and selectively grow out the spleen-to-myeloma hybrids. The

Table I. Marine Toxins[1]

Toxin	Source	Dose and route	Mechanism of action	Comment
Cephalotoxin	Octopus vulgaris	150-300 ug subcutaneous (in the dog)	Lowers blood pressure, arrests heart and respiration; neurotoxicologically similar to saxitoxin, tetrodotoxin	Human deaths have occurred from octopus bites
Ciguatoxin	Coral reef organisms		Neurotoxic effects; nausea, vomiting, diarrhea; recovery after several weeks	Occurs in moray eel, barracuda, several fish species off Ryukyu Islands
Halitoxin	Haliclona viridis	270 mg interperitoneal (in the mouse)	Blocks nerve-muscle junction; effective against cancer cells	
Holotoxin	Holothuria tubulosa	5-15 mg intravenous (mouse)	Cardiotoxic	A toxin of the sea cucumber
Nereistoxin	Lumbriconereis heteropoda	33 mg intravenous (mouse)	Causes muscle paralysis, halts respiration	Used as an insecticide in Japan

Palytoxin	Palythoa spp.	100 ug intraperitoneal (mouse)	Constricts coronary artery	Effect in man occurs upon ingestion of the file fish (Alutera)
Saxitoxin	Saxidomus giganteus	10 ug intraperitoneal (mouse)	Causes neuromuscular junction paralysis	Effect in man occurs upon ingestion of clams and mussels
Tetrodotoxin	Puffer fish	8-20 ug gastrointestinal (mouse)	Neurotoxic: causes respiratory paralysis	Used in clinical trials for pain in neurogenic leprosy

[1]The dose per kilogram is based on LD_{50}, the lethal dose required to kill 50 percent of a given animal population. From Kaul (1979)

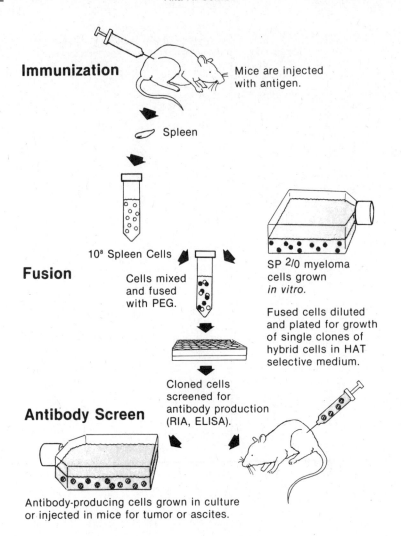

Immunization — Mice are injected with antigen.

Spleen

10^8 Spleen Cells

Fusion — Cells mixed and fused with PEG.

SP 2/0 myeloma cells grown *in vitro.*

Fused cells diluted and plated for growth of single clones of hybrid cells in HAT selective medium.

Cloned cells screened for antibody production (RIA, ELISA).

Antibody Screen

Antibody-producing cells grown in culture or injected in mice for tumor or ascites.

Figure 1. Procedure for production of monoclonal antibody-producing cells.

cultured myeloma cell possesses a mutation which results in a lack of the enzyme hypoxanthine phosphoribosyl transferase (HPRT), rendering it sensitive to killing by aminopterin. Spleen cells, however, contain normal HPRT, allowing spleen-to-myeloma hybrids to survive. The selective medium does not kill normal spleen cells, but these do poorly in culture and either divide slowly or die spontaneously. The cell mixture, with selective medium, is plated out in microtiter dishes at concentrations yielding one viable hybridoma in every second or

third well. Thus, the hybrids are cloned and each represents
the result of the fusion between a single myeloma cell and a
single spleen cell. In general, approximately 500 hybrids are
obtained from an immunized spleen. Most either do not make
antibody or make an antibody against an irrelevant antigen.
With a well-immunized donor animal, as high as 30 to 40 percent
of the hybridomas will make antigen reacting with the
immunogen. Of course, in most cases, only a few percent of the
surviving hybrids produce the desired antibody. The medium from
the wells in the dishes containing a viable hybridoma is
screened for antibody soon after the original fusion has been
done. Those hybrids producing antibody are grown into mass
culture and stored frozen. Also, cells are usually injected
into recipient mice, where they produce tumors and induce 10 to
15 ml of ascites fluid in each tumor-bearing mouse. Approximately 10 to 100 ug monoclonal antibody/ml is obtained in tissue
culture, but in ascites fluid as much as 10 mg antibody/ml can
be produced. The beauty of this technology is that all the
progeny of the hybridoma produce the same antibody molecule so
that the antibody is monoclonal and homogeneous (Scharff and
Roberts 1981).

Further applications of the technology in marine pharmacology are practically unlimited, including study of the structure
and function of the toxins, as well as for production of
anti-toxins for treatment. For example, this technology was
applied in our laboratory recently to produce monoclonal
antibodies to the enterotoxin of <u>Vibrio cholerae</u>, a brackish
water and estuarine bacterium which causes cholera.

Monoclonal antibodies against cholera toxin were produced to
obtain highly specific antisera to cholera toxin. Fifteen
hybridoma cell lines producing monoclonal antibodies specific
for the determinants of cholera toxin were derived from the
fusion of mouse myeloma cells and spleen cells from mice
immunized with cholera toxin (see Figures 2 and 3). The cell
lines were stabilized, examined for specific antibody production, and immortalized by freezing cultured cells and tumor
cells which had been grown subcutaneously in mice. All of the
cell lines continued antibody secretion upon thawing. The
antibodies produced by the hybridoma cell lines were characterized by determination of the class of light and heavy chain
components and by determination of specificity for cholera toxin
subunit. All of the antibodies contained the kappa light chain,
four contained the u heavy chain, and the remaining 11 the G_1
heavy chain. Ten of the monoclonal antibodies were specific for
the B subunit of cholera toxin and three were specific for the A
subunit. The remaining two appear to react with both subunits
(Figure 4) (Remmers et al. 1982).

Molecular and genetic studies have shown that possession of
the 01 serological antigen of <u>Vibrio cholerae</u> is not correlated
with possession of toxin genes (Kaper et al. 1981). Thus, both
non-01 and 01 strains of <u>Vibrio cholerae</u> can be potentially

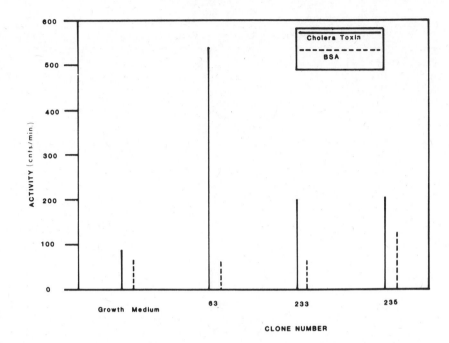

Figure 2.   Specificity of three stabilized clones derived
from hybridization, using primed mouse spleens.   Bar height
indicates activity of bound ^{125}I-labeled sheep anti-mouse
immunoglobulin.   The solid bars indicate activity against
cholera toxin-coated plates.   The broken bars indicate activity
against the unrelated antigen, BSA, coated plates.

toxigenic.   Furthermore, vibrios--including various strains of
Vibrio cholerae--may produce functionally and/or biochemically
related toxin molecules.   Since monoclonal antibodies detect
specific antigenic determinants, a battery of monoclonal
antibodies will be helpful in evaluating the antigenic related-
ness of toxins elaborated by various clinical and environmental
vibrio strains.   Subunits A and B of the cholera toxin were able
to be distinguished by the monoclonals specific for the A or B
antigen.   Clearly, characterizing the elusive ciguatera and
related toxins by this method will provide needed knowledge for
detection and possible treatment or protection.
     Halitoxin, a toxic complex of several marine sponges of the
genus Haliclona, has been isolated, partially purified,
spectrally characterized, and chemically degraded, yielding a
proposed chemical structure for the toxin (Schmitz et al.
1978).   The toxin has proved to be a complex mixture of high
molecular weight and toxic pyridinium salts, and can be isolated
from the sponges, Haliclona rubens, H. viridis, and H. erina.
The sponge extracts are toxic for fish and mice (LD_{50} of

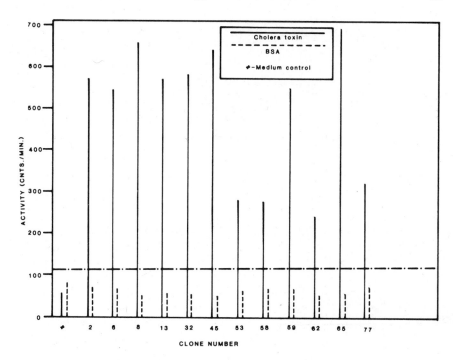

Figure 3. Specificity of twelve stablized clones derived from hybridization using multiple boosted mouse spleens. See legend, Figure 2. The horizontal line indicates a signal-to-noise ratio of 2.

approximately 275 mg/kg) and also inhibit the growth of Ehrlich ascites tumors (Baslow 1969; Turlapaty et al. 1973). Thus, the halitoxin(s) may prove to be an anti-tumor agent or agents. The hybridoma approach to characterizing these compounds should permit wider screening for the toxin amongst a variety of sponges as well as for subsequent purification and testing.

Similarly, lophotoxin, a new neuromuscular toxin isolated from several Pacific gorgonians of the genus Lophogorgia, has been isolated and purified (Fenical et al., 1981). Originally discovered during a search for chemical defense adaptations of marine organisms, a variety of horny corals or gorgonians (sea fans and whips; phylum Cnidaria, order Gorgonaceae) in tropical or subtropical waters were studied; and cytotoxic, ichthyotoxic and antibacterial activity was noted. Lophotoxin inhibits nerve-stimulated contraction without affecting contraction evoked by direct electrical stimulation of the muscle. The data suggest that epoxylactone and furanoaldehyde groups may be responsible for the potent biological properties of lophotoxin.

Palytoxin, an extremely poisonous, water-soluble substance

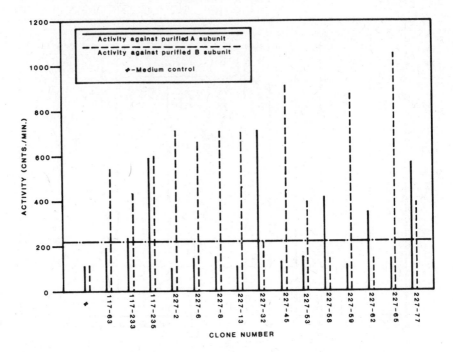

Figure 4. Subunit specificity of monoclonal antibodies
prepared against cholera toxin. The solid bars indicate
activity against purified A subunit. The broken bars indicate
activity against purified B subunit. The horizontal line
represents a signal-to-noise ratio of 2.

from marine coelenterates belonging to the genus Palythoa (Moore
and Scheuer 1971), has recently been described and its structure
elucidated (Moore et al. 1980, 1982; Moore and Bartolini 1981;
Fox, 1982). The toxic nature of zoanthids of the genus Palythoa
was discovered by Ciereszko and Attaway in 1961 (Attaway 1968).
The component responsible for the toxicity, palytoxin (PTX), was
isolated several years later independently both at the
University of Oklahoma (Attaway and Ciereszko 1974) and the
University of Hawaii (Moore and Scheuer 1971). Palytoxin was
used by natives on Maui to tip their weapons as a defensive
advantage against invaders from the island of Hawaii.

The molecular weight of the toxin is 2,681 daltons, with
three nitrogens in the molecule (Moore et al. 1980), and an
elemental composition of $C_{129} H_{223} N_3 O_{54}$. The complete
stereochemical structure has been described, with the toxin
concluded to be an entirely new class of natural products. Its
biogenesis is not at all clear, but it appears to be modified
B-alanine units attached to a highly functionalized fatty acid
chain that is 115 carbons long (Moore et al. 1982).

Palytoxin contains no repeating units, as is common in medium-sized molecules in nature. It is neither a peptide nor a terpene, but, rather, more like the polyether antibiotics. Some of the nonadjacent carbons along the chain are interconnected by ether, ketal, and hemiketal linkages, with no carbocyclic rings in the molecule. An organic chemist, Yoshito Kishi, and his colleagues at Harvard University are working on the synthesis of palytoxin. Palytoxin is very likely the most toxic nonprotein known, similar to ricin in potency, a polypeptide toxin found in castor beans. Although not quite as deadly as the toxin of Clostridium botulinum, it exerts a lethal effect when administered intravenously in low concentrations to lab animals.

Palytoxin influences calcium and potassium ion transport in nerves and the heart. Animals undergo paralysis and heart failure. The mechanism of the cardiovascular effects of palytoxin has been reviewed by Kulkarni et al. (1978).

The fascinating aspect of palytoxin is that it is synthesized by a marine Vibrio sp. growing symbiotically with the coelenterate Palythoa and apparently related to Vibrio cholerae. This vibrio reportedly is only mildly pathogenic for humans (Fox 1982; Moore et al. 1982), causing a flu-like illness, and is rapidly attenuated in laboratory culture, losing its ability to make toxin. The Palythoa sp. containing the toxin grows in a tide pool along the coast of Maui not far from Hana. By land, the tide pool is virtually inaccessible and native fishermen avoid that part of the coast (Fox 1982; Moore et al. 1982).

Clearly, the palytoxin represents an excellent candidate for cloning of the gene(s) synthesizing the toxin. The general principles of gene manipulation should apply (Old and Primrose 1980), and production of the toxin in quantities large enough for structural analysis and pharmacological studies should, thus, be possible, allowing development of a treatment, as well as analysis of the mode of action of the toxin.

The need, at the moment, is for strategies for collecting, culturing, and screening marine organisms from which bioactive agents can be isolated and characterized. Most likely, the immediate successes will occur in discoveries of novel antibacterials or antibiotics produced by marine bacteria. However, the potential for engineering the production of the more complex pharmaceuticals exists. Ingenuity will certainly provide the means and the initiative.

Industrial Chemicals

Marine toxins are fascinating from a scientific point of view, but it is more likely in the short term that the marketable products will come from marine polysaccharides, carotenoids, and specialty chemicals such as unusual sugars, enzymes, and algal lipids. These represent products for short-term and rather immediate pay-off.

In fact, carrageenan is a major product from the red seaweeds and is widely used as an extender in foods and related
products from evaporated milk to toothpaste. Agarose is widely
employed in electrophoresis and chromatography analyses in the
laboratory. Thus, seaweed culture offers an opportunity for
gene cloning and transfer in microbial processes that can extend
the presently profitable market.

Specialty chemicals from salt-tolerant microbial systems,
notably polysaccharides, enzymes, and lipids offer the greatest
potential in the immediate future.

Besides toxins and biologically active substances and those
substances already exploited commercially, such as carrageenan,
chitin, and agarose, a variety of interesting compounds and
metabolites have also been reported, including spatane
diterpenoids, from the tropical marine alga Stoechospermum
marginatum, not yet observed from terrestrial sources (Gerwick
et al. 1981).

A new polyether derivative of a C_{38} fatty acid, okadaic
acid, has been isolated from Halichondria (syn. Reniera) okadai
Kadota, a black sponge commonly found along the Pacific coast of
Japan, and H. melanodocia, a Caribbean sponge found in the
Florida Keys (Tachibana et al. 1981). The structural features
of okadaic acid suggest that it belongs to the class of compounds known as ionophores, hitherto known only from terrestrial
microorganisms. It is likely that okadaic acid is a metabolite
of an epiphytic microorganism, rather than of Halichondria spp.
Also, however, structural studies of the marine toxins palytoxin
and ciguatoxin strongly indicate that the two compounds possess
many structural features normally associated with ionophores.
In fact, T. Yamusoto has isolated okadaic acid from the
dinoflagellate Procentrum lima, a likely progenitor of okadaic
acid (Tachibana et al. 1981).

Sponges and gorgonians have been useful sources of
biologically active metabolites because they are frequently
abundant, permitting pursuit of trace metabolites. These
unusual compounds may be pathway intermediates and also offer
potential sources of new chemicals. An antimicrobial
sesterterpene 1, called palauolide, was recently described by
Sullivan and Faulkner (1982).

Sea hares offer the advantage of being rich sources of
interesting metabolites, but the ultimate source of the latter
is not always clear, often proving to be algae. Recent work on
pharmacologically active substances associated with sea hares,
coelenterates, and sponges has been summarized by Schmitz et al.
(1981). Notable are the extracts of the sea hare Aplysia
dactylomela which show both cytotoxicity and in vivo antitumor
activity, as well as a variety of halogenated metabolites.

Recently the cloning and expression of sea urchin histone
genes using SV-40 DNA as a vector have been reported (Kapstein
and Fareed 1979). Thus, the cloning of genes of marine animals

has begun, but the important future opportunities remain to be
grasped.

Microbiologists have long been familiar with the diversity
of microorganisms. More than 50,000 defined species have been
described, and the widespread occurrence of these microorganisms
in soils, at concentrations of millions per gram, has been
exploited in the production of widely used medical products of
microbial origin. Greatest attention in the past has been paid
to natural products with antibiotic properties. Yet natural
products occur in fermentation broths during secondary metabolism, a characteristic of the incomplete metabolic control
operative in growth-inhibited microorganisms. This general
mechanism of biosynthesis of natural products by microorganisms
can be exploited very effectively as shown by the following.
The top ten commercial products in a list of pharmaceutical
products sold in Japan, ranked in order of market size, revealed
four with a total market value exceeding 300 million dollars per
year which were derived directly from microorganisms. Furthermore, these four products were sold for purposes other than
antibiotic action. Five others listed (see Table II) are
antibacterials, produced as chemical modification of
Cephalosporin C. Only one of the ten was discovered by total
synthesis. In fact, microorganisms do not preferentially
synthesize only antibacterials, but, rather, a potpourri of
natural substances, many of which, originating from biochemical
pathways, can influence the biological activities of higher
forms of life (Woodruff 1980).

Secondary metabolites produced in association with other
life forms which are experiencing a shift from trophophase to
idiophase (as defined by Bu'Lock [1965, 1967]), may experience
rapid multiplication to large numbers followed by abrupt
cessation of growth, even though supplies of most nutrients
remain abundant and environmental conditions remain favorable.
In trophophase, the demand is for metabolites as basic cell
constituents: amino acids, Krebs-cycle intermediates, and
purines. At idiophase, the cells alter when replication ceases,
as in commercial fermentation vessels, because of nutrient
deficiency, oxygen depletion, toxic product accumulation, or
other limiting factors. Despite enzyme repression and feedback
inhibition of anabolic pathways, metabolic intermediates
continue to accumulate, serving as initiators for secondary
metabolism, yielding products that can be pharmaceutically
useful. No doubt these interactions occur in the environment,
with accumulation of natural products, for example, in the sea
hare and in marine bacteria found in association with marine
animals and plants or free-living in the water column or
sediments of the world oceans. The message is, clearly, that
directed search for non-antibiotically active natural products
in the marine environment, especially those described earlier as

Table II. Ranking of pharmaceutical products in Japan on
the basis of the sales volume between August 1, 1977, and July
31, 1978; NP, natural products

Rank	Product	Type	Source
1	Cefamezin	Antibiotic	NP: semisynthetic
2	Keflin	Antibiotic	NP: semisynthetic
3	Keflex	Antibiotic	NP: semisynthetic
4	Futraful	Antitumor	Synthetic
5	Krestin	Immune enhancer	NP
6	Neuquinone	Coenzyme	NP: also semi-synthetic
7	Larixin	Antiobiotic	NP: semisynthetic
8	Syncl	Antibiotic	NP: semisynthetic
9	Dasen	Enzyme	NP
10	Picibanil	Immune enhancer	NP

From Woodruff (1980)

being unique to marine organisms, can open an entirely new
source of industrial chemicals. What is needed are new and
novel screening strategies for such products. Then, the
techniques of genetic engineering will remove the limitations of
remote geography and the need for large harvests. Cloning genes
effective for producing desired compounds in a non-marine,
industrially adapted vector offers rich sources of new products.

Biodegradation in the Marine Environment

In contrast to natural products, human-made compounds are
relatively refractory to biodegradation, often because organisms
naturally present cannot produce enzymes necessary for transformation of the original compounds, such that the resulting
intermediates can enter into common metabolic pathways to be
metabolized completely, and this creates special problems for
waste treatment and environmental protection.

Required steps to initiate biodegradation are reasonably well understood. Halogenated compounds, we know, are particularly persistent because of the location of the halogen atom, the halide involved, and the extent of halogenation. Groups of microorganisms useful for treating specific types of human-made compounds have been compiled by Kobayashi and Rittman (1982). Selective use of microorganisms, including actinomycetes, fungi, bacteria, phototrophic microorganisms, anaerobic bacteria, and oligotrophic bacteria, is not new and represents a common practice in certain applications, such as wastewater treatment for biological removal of nitrogen via sequential nitrification and denitrification. Controlled mixed cultures are already in use in Japan for treating selected industrial wastes in reactors. These cultures are composed of heterotrophic bacteria, photosynthetic bacteria, and algae (Kobayashi and Rittman 1982). Various methods of genetic engineering will certainly become widely used to develop optimized proliferation and maintenance of selected populations.

What has not yet been considered, however, is the engineering of microorganisms to be added to wastes that are to be discharged into the marine environment. It is obvious that, with increased use of the world oceans for man's waste, attention must be paid to the problems of marine pollution. Pollutants entering the ocean that can interfere with the integrity of ecosystems include synthetic organics, chlorination products, dredged spoils, litter, artificial radionuclides, trace metals, and fossil fuel compounds. Toxaphene, a group of slightly under 200 compounds produced by chlorination of wood waste products and camphors under ultraviolet light, contains carcinogenic and mutagenic members and may be more persistent in the environment than DDT and its degradation products (Goldberg 1981).

The problems of in situ degradation are much greater than for contained application. The modifications of genetic information resident in microorganisms that are useful in pollution control are: (1) amplification of enzyme concentrations in an organism, either by selection of constitutive mutants, increase in the number of copies of the gene for the enzyme, or both; (2) rearrangement of regulatory mechanisms controlling the expression of specific genes in response to specific stimuli; (3) introduction of new enzymatic functions into organisms not possessing them; and (4) alteration of the characteristics of specific enzymes, such as substrate specificity, kinetic constants (K_m and V_{max}), or factors such as pH optimum. To achieve these modifications, one can undertake in vitro recombinant DNA manipulation; in vivo modification via transposon mutagenesis or other transposon-mediated gene manipulation; genetic exchange via transduction, transformation, or conjugation; protoplast fusion; specific site mutagenesis; and specialized selection procedures to enrich for mutants. Again, what has not been considered to date is the

engineering of microorganisms capable of flourishing in the
marine optimized proliferation and maintenance of selected
populations.

From another perspective, the need for algicides and anti-
fouling agents is so great that breakthroughs in obtaining
compounds with these activities will guarantee market success.

In addition, specific problems of seafood industry wastes,
such as shellfish waste, have not been considered from the
viewpoint of engineering microorganisms to biodegrade the wastes
rapidly, even though conversion of the shellfish waste, chitin,
to single-cell protein has been considered (Revah-Moiseev and
Carroad 1981). Application of genetic engineering to improved
food yield is clearly a very promising area upon which to focus.

The industry developed to exploit chitin or its derivatives
remains small. Two companies in the United States produce
chitosan, and there is some production and marketing of chitosan
in Japan. This area of research, i.e., genetic engineering
applied to pollution technology, should follow quickly on the
heels of any breakthroughs occurring in waste treatment
processing.

Biotechnology Applications in Aquaculture

Bioengineering can bring big payoffs for aquaculture.
Marine microorganisms offer new sources of biomass and represent
a major opportunity for the future. In the United States, most
of the traditional fisheries are being harvested at or near
maximum sustainable yields. Approximately sixty percent of the
fisheries products consumed in the United States are imported,
representing a trade deficit in excess of $2.5 billion.
Mariculture offers the potential for reducing this deficit.
Exploiting microbial sources of protein at the larval stages and
during larval metamorphosis and growth can reap enormous
benefits and profit.

That aquaculture itself can pay off is already established.
China, for example, produces two million metric tons of finfish
every year, mostly in the form of carp grown in ponds, lakes,
reservoirs, and ditches. In the United States, interest in
aquaculture is on the rise, and that interest has meant
increases in our knowledge of marine biology and the technical
expertise to apply discoveries in marine biology to aquaculture. The United States has the opportunity to take the lead
in applying sophisticated, more efficient methods for fish
aquaculture, and Sea Grant has been at the forefront of many of
this country's advances in aquaculture. Research has been
underway for more than a decade on marine shrimp, freshwater
prawn, crawfish, blue crab, brine shrimp, salmon and other
finfish, oysters, clams, abalone, and scallops.

The salmon has long been a lucrative finfish along the
Pacific coast. Two types of aquaculture are used for raising
salmon--ocean ranching and pen-rearing--and both types pose
challenging research problems. Sea Grant projects underway at

The University of Alaska, the University of Washington, and the University of Idaho are using highly sophisticated methodologies--and a good deal of common sense--to solve some of the problems.

At the University of Alaska, researchers are looking at breeding between cultured and natural stocks to see if breeding takes place and to determine the effects and outcomes of any breeding that does. Alaska, as well as other universities, is working on developing a cheap feed for salmon--something that could make a big difference in the economics of closed-system aquaculture operations. Still another Alaska project is examining effects of temperature on salmon fry to assess--among other objectives--the best time to release salmon to avoid high mortalities.

At the University of Washington, a salmon stock enhancement program has been underway since 1977. Present projects focus on population dynamics, incubation effects on fry, and improved salmon nutrition. Researchers there are also attempting to develop a salmon broodstock selected for pen culture, that is, fish growing faster and healthier and bigger on minimal feed. Washington Sea Grant's salmon studies include analysis of the relation of the thyroid gland to smoltification, important since many salmon die before their adaptation to the sea, and an investigation into Vibrio anguillarum, a bacterial pathogen of fish which causes problems for mariculturists worldwide.

In an attempt to modify salmon behavior, researchers at Oregon State University are studying the dynamics of imprinting and the interrelationships between dietary lipids and protein and growth. Still other Sea Grant salmon studies are going on at the University of California at Berkeley, at Humboldt State University, and at the University of Idaho.

On the Atlantic coast, University of Maine Sea Grant salmon-related work is focusing on infectious pancreatic necrosis (IPN). There the antigenic relationships of various IPN isolates are being studied in order to develop polyvalent antisera and effective vaccines against the virus. Temperature-sensitive mutants of the virus have also been isolated to provide genetic information and possible sources of attenuated virus strains for use in live vaccines. The use of hybridoma technology will almost certainly pay off in the production of fish vaccines.

The Atlantic salmon, formerly plentiful in New England's major river systems, is the focus of salmon research at the University of Rhode Island, designed to decrease stress and disease, to increase viable eggs, and to produce a population suitable for culture.

In North Carolina, researchers are studying eel; at the University of Texas, red drum and red snapper; at Cornell, the walleye; at Woods Hole, sea-run brook trout.

The culture of bait organisms has also been achieved. At the University of Florida, the lugworm (Arenicola cristata) is

being cultured; and at the Hawaii Institute of Marine Biology,
investigators are concentrating on cost-effective culturing of
bait minnows. At the University of Minnesota, the focus is on
the bait leech. Thus the range of species considered for
culture is wide. Although some of this work is highly
experimental, much of it has practical applications.

A major problem of mariculture is disease, predominantly
microbially mediated infections and epidemics. Viral and
bacterial agents that are common hatchery complaints include
IPN, as noted above, egtved and other viruses, and Vibrio and
Aeromonas among the bacteria (Ahne 1980). Many causes of
disease and loss of hatchery stocks are still not yet known, nor
are controls of epizootics yet available.

An extract from Ecteinascidia turbinada (Ete) has been shown
to enhance the hemocyte function of invertebrates, e.g., blue
crabs (Callinectes sapidus), crayfish (Procambarus clarkii), and
prawns (Macrobrachium rosenbergii), possibly rendering the
animals more resistant to infection. Interestingly, intraperi-
toneal injection of Ete renders eel strongly resistant to
Aeromonas hydrophila and appears to create the potential for
phagocytic activity. Ete also causes changes in the concentration of peripheral blood leucocytes (Sigel et al. 1970; Sigel
1974; Lichter et al. 1975; Lichter et al. 1976; Sigel et al.
1982).

Biotechnology offers extraordinary opportunities for aquaculture. Many species of shellfish and finfish are available in
culture, providing excellent opportunities for selection and
gene manipulation. Production, stabilization, and delivery of
vaccines, employing both hybridoma technology and genetic
engineering, will enhance productivity from the egg through the
larval stages, presently a high-risk portion of the life cycle.

Stock assessment of migrating fish and species
identification remain unresolved issues in fishery management.
A method for comparison of mitochondrial DNA (mt DNA) from
different individuals offers an opportunity for mapping the mt
DNA genome and is being explored by several investigators
(Power, personal communication). The use of restriction
endonucleases, which cleave DNA at sites specified by unique
sequences of four, five, or six nucleotides, allows mt DNA, when
digested by one of the enzymes, to be cut into a characteristic
set of fragments of different lengths (Boyer 1974; Brown and
Vinograd 1974). The number and sizes of the fragments can be
related to mt DNA sequence homology when compared with that of
other individuals. The method of using restriction digests to
analyze mt DNA has permitted investigation of evolutionary
relationships among species and conspecific populations. When
applied to marine fishes and invertebrates, analysis of mt DNA
offers a valuable mechanism for assessing the population
structure of marine fishes. Genetic differences among fish
species have been successfully detected using gel electro-

phoresis of proteins in tissue extracts (Gharrett and Utter 1982).

The possibility of using genetic marking, by introducing selected genetic traits into fish, as opposed to mechanical tags, is yet another avenue for genetic engineering in aquaculture. Fish wander over long distances without barriers to their movements, but there may now be a way to detect subpopulations or specific independent stocks by very precise methods yielding unequivocal results.

Marine plants also offer special opportunities, and genetic engineering of osmoregulation, for example, is being studied (Rains and Valentine 1930). Plants which are halophytes can be useful in agricultural areas where the soil has become too salty for conventional agriculture, and application to marine and estuarine grasses can prove beneficial in managing erosion and shoreline losses.

As a source of new and larger biomass, the oceans are clearly under-exploited, for various technological reasons, but genetic engineering offers significant opportunities for new sources of biomass for food.

Biofouling
Fouling of surfaces in the marine environment is a costly burden for any operation carried out in or near seawater. The progressive steps involved in biofouling, from the initial primary film to attachment of invertebrate animals capable of boring and digesting the surface (Smedes and Hurd 1981), have been documented in countless publications during the last century. The ability of bacteria to find, attach, adhere, elaborate specific film-forming substances, and regulate expression of all of these functions is fundamental to the fouling process. Fortunately, the tools of genetic engineering are well-suited to analysis of the properties of bacterial cell surface components, since specific genetic elements determining the structure of specific polysaccharides and polypeptides can be isolated and the mechanism of cell-surface association probed so that appropriate steps for intervention can be taken.

There is, however, another facet of the biofouling issue, a positive side. Larvae of invertebrates have been shown to prefer to settle on surfaces coated with microbial films (see Table III). Graham et al. (1980) have shown that settlement and metamorphosis do not occur in the absence of microbial films in the case of the tube-forming polychaete, Janua (Dexiospira) brasiliensis. Janua is a small (2 to 3 mm), hermaphroditic polychaete abundant on a variety of surfaces, notably Zostera (eel grass).

Planula larvae of the medusa Cassiopeia andromeda settle on a substrate, attach, form a pedal disc and metamorphose, i.e., elongate, segregate stalk and calyx, and develop a hypostome and tentacle anlagen (Neumann 1979). Pedal-disc formation and

Table III. Chemical inducers of specific settlement of invertebrate larvae

Invertebrate	Inducer	Reference
Barnacles	Arthropodins (proteins)	Crisp, 1974
Coelenterates	Iodinated proteins Lag phase bacterial products	Spangenberg, 1971 Bourdillon, 1954 Muller, 1973
Sponges	Pseudomonas lectin	Muller, et al., 1981
Oysters	Iodinated proteins Thyroxine L–Dopa D–Dopa/Melanin Isochrysis	Veitch and Hidu, 1971 Veitch and Hidu, 1971 Coon, 1980 Weiner and Colwell, 1982 Mitchell and Young, 1972
Bivalves	GABA Algal Phycoerythrobilins Coral mucus Chlorine compounds	Morse, et al., 1980 Morse, et al., 1980 Hadfield, 1978 Bonar, 1974 Hadfield, 1978
Sea Urchin	Low MW bacterial byproduce	Cameron and Hinegardner, 1974

metamorphosis were clearly shown to be initiated by substance(s) produced by a marine Vibrio sp. The inducing factor is released into the culture medium by the bacteria.

For commercially important shellfish, such as the American oyster Crassostrea virginica, enhanced larval set and metamorphosis can be critical to a successful hatchery operation. Thus, a nuisance for one individual, viz., biofouling of a racing yacht, can be welcomed by another, the hatchery operator wishing to control spat set of oysters.

A few years ago, a unique bacterium, which we call LST, was isolated from tanks containing oyster larvae at the mariculture unit in Lewes, Delaware. It appears to be a member of a new genus, since DNA prepared from this isolate does not hybridize with any other Vibrio DNA's. This organism has since been reisolated from oysters and oyster beds. Interestingly, it is associated with induction of settlement and metamorphosis of the

oyster. The strain adheres very strongly to cultch and other
hard surfaces, forming microcolonies on the cultch. In
sufficient numbers, notably during the decline phase of growth,
the bacterium produces a high concentration of pigment,
sufficient to attract oyster larvae, and demonstrates a
hormone-like, stimulatory effect on larval development and
metamorphosis. Such a symbiotic relationship between bacteria
and an invertebrate is common in the marine environment, as
with, for example, the relationship between the sponge
Halichondria panicea and Pseudomonas insolita.

Work done in collaboration with Dr. Ronald Weiner at the
University of Maryland has revealed the production of melanin
(and its precursors), with a brown pigment produced by the
bacterial strain LST, which acts as a metamorphosis-inducing
agent for oyster larvae.

From chemical and physical criteria, the pigment has been
identified as melanin. The pigment has been concentrated and
partly purified by a series of centrifugations, precipitations,
and dialysis procedures. The molecular weights of the pigment
and precursors are 12,000 to 120,000 daltons.

Research accomplished during the last fifty years has made
it evident that planktonic larvae of benthic invertebrates
settle and metamorphose in response to specific substances or
conditions in their environment. In the absence of those
substances, metamorphosis can be delayed indefinitely.

In only a few cases has the metamorphosis-stimulating
substance been isolated and identified. Barnacle larvae will
metamorphose in response to contact with a species-specific
arthropodin, a soluble, tanned protein present in the cuticle of
adult barnacles. Iodinated proteins are active on coelenterates
(anthozoan planulae and scyphozoan scyphystomae). The same
class of compounds has been isolated from the mantle cavity of
adult oysters and stimulates metamorphosis of oyster larvae.
Bivalves and gastropods (larvae of the pink abalone) can be
induced to metamorphose in the presence of gamma-amino butyric
acid (GABA). Bacteria mediate metamorphosis of coelenterates
(Hydractinea planulae larvae) and the sea urchin--in the latter
case, in association with a 1-5 kilodalton protein. A
carbohydrate produced by Pseudomonas marina stimulates
invertebrate development via a lectin interaction (R. Mitchell,
personal communication). Thus, specific chemical inducers
trigger invertebrate morphogenesis. These inducers, as in the
case of melanin, are often amino acids and/or pigments produced
by bacteria.

Recently, D.E. Morse, professor of molecular genetics and
biochemistry, and research specialists N. Hooker and A. Morse
devised techniques to help control reproduction, larval
development, metamorphosis, survival, and growth in a species of
abalone, for which they received recognition at the
International Symposium of the World Mariculture Society in
Venice, Italy. A naturally occurring chemical causing abalone

to metamorphose has been described by these workers, opening the
door to genetic engineering aimed at cloning specific genes to
enhance growth of the abalone.

Thus, specific compounds of bacterial origin--for example,
those related to melanin--can promote settlement and development
of American oysters and very likely of other marine invertebrates as well, including those of commercial value.

Another fascinating aspect of the molecular biology of
marine invertebrates that is presently being studied is
protochordate allorecognition. Colonial tunicates, unlike
vertebrates, undergo transplantation in nature. Rejection or
acceptance of colonies has been shown recently by Scofield et
al. (1982) to be controlled by a single gene locus with multiple
alleles. The same genetic region apparently maintains this
polymorphism by preventing fertilization between gametes sharing
alleles. Thus, the histocompatibility system of marine invertebrates can provide a better understanding of the immune
recognition of tissue antigens and rejection of transplanted
allogeneic organs, tissues, and cells in vertebrates, involving
the major histocompatibility complex (MHC), since the
protochordate allorecognition is controlled by an MHC-like gene
system (Scofield et al. 1982), a sort of ancestral MHC gene
complex (Scofield and Weissman 1981).

Conclusion

The few examples offered here illustrate an immense
potential of biotechnology for the marine sciences, but they
scarcely begin to reveal opportunities which lie ahead. Because
of a decade of research supported by the National Sea Grant
College Program, there exists a strong foundation for
exploitation of biologically active compounds already known to
occur in the sea and encouragement for further exploration into
the recesses of the world's oceans for compounds and food
sources as yet undiscovered.

The greatest opportunity of all is represented by the
applications of genetic engineering to the marine sciences. We
will have access now to an untapped gene pool representing
transport systems for minerals; metal concentration; novel
photosynthetic systems; marine pheromones, i.e., "communicator
substances" produced by marine animals; as well as the hydrogen
sulfide utilizing and microbially mediated system of the
Galapagos Vent ecosystems. A wealth of genetic information can
now be tapped and new knowledge of ourselves and of our oceans
can be gained.

The potential of the oceans to serve as a significant source
of protein on an even grander scale will come closer to being
realized with the advent of the tools of genetic engineering.
The scale of management and stock breeding for domestic animals
can now be duplicated for fish and shellfish--providing an even
greater advantage, compared with domestic livestock, since the

generation time and maturation cycle is significantly shorter for marine species.

The ultimate promise of the oceans is, in a sense, the transmogrification of humankind. Soon we shall learn what more wondrous gifts the seas will offer.

References

1. Ahne, W., 1980. Fish Diseases, Springer-Verlag, New York, 252 pp.

2. Attaway, D.H., 1968. Ph.D. Dissertation, The University of Oklahoma, Norman, OK.

3. Attaway, D.H. and L.S. Ciereszko, 1974. Isolation and partial characterization of Caribbean palytoxin, In Proc. Sec. Internat. Coral Reef Symp., Brisbane, Australia, p. 497.

4. Austin, B., D.A. Allen, A. Zachary, M.R. Belas and R.R. Colwell, 1979. Ecology and taxonomy of bacteria attaching to wood surfaces in a tropical harbor, Can. J. Microbiol., 25:447-461.

5. Baslow, M.H., 1969. Marine Pharmacology, Williams and Wilkins, Baltimore, 273 pp. Republished with update, Robert E. Krieger Publishing Co., Huntington, N.Y., 1977.

6. Bonar, D.B. and M.G. Hadfield, 1974. Metamorphosis of the marine gastropod _Phestilla sibagae_: I. Light and electron microscopic analysis of larval and metamorphic stages, J. Exp. Mar. Biol. Ecol., 16:227-255.

7. Bourdillon, A., 1954. Mise en evidence d'une substance favorisant la metamorphose des larvae d'_Alcyonium coralloides_ (Van Koch), Comptes Rendues Acad. Sci., Paris, 239:1434-1436.

8. Boyer, H.W., 1974. Restriction and modification of DNA: Enzymes and substrates, Fed. Proc., 33:1125-1127.

9. Brown, W.M. and J. Vinograd, 1974. Restriction endonuclease cleavage maps of animal mitochondrial DNAs, Proc. Nat. Acad. Sci., 71:4617-4621.

10. Bu'Lock, J.D., 1965. The Biosynthesis of Natural Products: An Introduction to Secondary Metabolism, McGraw-Hill, New York.

11. Bu'Lock, J.D., 1967. Essays in biosynthesis and microbial development, Wiley, New York.

12. Cameron, R.A. and R.T. Hinegardner, 1974. Initiative of metamorphosis in laboratory cultured sea urchins, Biol. Bull., 146:335-342.

13. Coon, S., 1980. Personal communication.

14. Crisp, D.J., 1974. Factors influencing the settlement of marine invertebrate larvae, In P.T. Grant and A.M. Mackie, eds, Chemoreception in Marine Organisms, Academic Press, New York, pp. 177-265.

15. Elston, R., E.L. Elliot and R.R. Colwell, 1981. Shell fragility, growth depression and mortality of juvenile American and European oysters (Crassostrea virginica and Ostrea edulis) and hard clams (Mercenaria mercenaria) associated with surface coating Vibrio spp., bacteria, Proc. Nat. Shellfish Assn. (Abstr.)

16. Fenical, W., R.K. Okuda, M.M. Bandurraga, P. Culver and R.S. Jacobs, 1981. Lophotoxin: A novel neuromuscular toxin from Pacific sea whips of the genus Lophogorgia, Science, 212:1512-1514.

17. Fox, J.L, 1982. Complex structure of marine toxin unraveled, In Chem. Eng. News, Jan. 4, 1982, pp. 19-20.

18. Gerwick, W.H., W. Fenical and M.U.S. Sultanbawa, 1981. Spatane diterpenoids from the tropical marine alga Stoechospermum marginatum (Dictyotaceae), J. Organ. Chem., 46:2233-2241.

19. Gharrett, A.J. and F.M. Utter, 1982. Scientists detect genetic differences, Sea Grant Today, 12:3-4.

20. Goldberg, E.D., 1981. The Crystal Mountain report: An approach to defining ocean assimilative capacity, In Use of the Ocean for Man's Wastes, Symposium Proc. June 23-24, 1981, National Academy Press, Washington, DC, pp. 1-10.

21. Graham, S., D. Kirchman and R. Mitchell, 1980. Larval settlement on microbial films: A model system, Biol. Bull., 159:160.

22. Gudger, E.W., 1930. Poisonous fishes and fish poisonings, with special reference to ciguatoxin in the West Indies, Am. J. Trop. Med., 10:43-55.

23. Hadfield, M.G., 1978. Metamorphosis in marine molluscan larvae: An analysis of stimulus and response, In F.-S. Chia, and M.E. Rice, eds., Settlement and Metamorphosis of Marine Invertebrate Larvae, Elsevier, New York, pp. 165-176.

24. Halstead, B.W., 1965. Poisonous and Venomous Marine Animals of the World, Vol. 1, U.S. Govt. Printing Office, Washington, DC.

25. Halstead, B.W., 1967. Poisonous and Venomous Marine Animals of the World, Vol. 2., U.S. Govt. Printing Office, Washington, DC.

26. Halstead, B.W., 1970. Poisonous and Venomous Marine Animals of the World, Vol. 3, U.S. Govt. Printing Office, Washington, DC.

27. Hashimoto, Y., 1979. Marine Toxins and Other Bioactive Marine Metabolites, Japan Scientific Societies Press, Tokyo, 369 pp.

28. Hollenbeak, K.H., F.J. Schmitz and P.N. Kaul, 1976. Cardiotonic agents from marine sponges: Isolation of histamine and N-methylated histamines, In Proc. Food and Drugs from the Sea Conference, Washington, DC, Marine Technol. Soc., pp. 282-287

29. Kaper, J.B., S.L. Mosely and S. Falkow., 1981. Molecular characterization of environmental and non-toxigenic strains of Vibrio cholerae, Infect. Immun., 32:661-667.

30. Kapstein, J.S. and G.C. Fareed, 1979. Cloning and expression of sea urchin histone genes using SV-40 DNA as a vector, J. Supramol. Struct., 8(3):7.

31. Kaul, P.N., 1979. The sea's biomedical potential. Impact of Science on Society, 29:123-134.

32. Kaul, P.N., 1981. Compounds from the sea with actions on the cardiovascular and central nervous systems, Federation Proc., 40:10-14.

33. Kaul, P.N. and C.J. Sindermann, 1978. Drugs and Food from the Sea: Myth or Reality?, University of Oklahoma, Norman, OK, 448 pp.

34. Kobayashi, H. and B.E. Rittman, 1982. Microbial removal of hazardous organic compounds, Environ. Sci. Technol., 16:170A-183A.

35. Kohler, G. and C. Milstein, 1975. Continuous cultures of fused cells secreting antibody of predefined specificity, Nature, 256:495-497.

36. Kulkarni, S.K., W.G. Kirlin and P.K. Kaul, 1978. Mechanism of cardiovascular effects of palytoxin, In P.N. Kaul and

C.J. Sindermann, eds., Drugs and Food from the Sea,
University of Oklahoma, Norman, OK, pp. 73-80.

37. Lawrence, D.N., M. Enriquez, R.M. Lumish and A. Maces,
    1980. Ciguatera fish poisonings in Miami, J. Amer. Med.
    Assoc., 244:254-258.

38. Lichter, W., A. Ghaffar, L.L. Wellham and M.M. Sigel, 1978.
    Immunomodulation by extract of Ecteinascidia turbinata, In
    P.N. Kaul and C.J. Sindermann, eds., Drugs and Food from the
    Sea: Myth or Reality?, University of Oklahoma Press,
    Norman, OK, pp. 137-144.

39. Lichter, W., D.M. Lopez, L.L. Wellham and M.M. Sigel, 1975.
    Ecteinascidia turbinata extracts inhibit DNA synthesis in
    lymphocytes after mitogenic stimulation by lectins, Proc.
    Soc. Exp. Med. Biol., 150:475-478.

40. Lichter, W., M.M. Sigel, D.M. Lopez and L.L. Wellham, 1976.
    Inhibition of DNA synthesis by Ecteinascidia turbinata
    extract (Ete), In H.H. Webber and G.D. Ruggieri, eds.,
    Food-Drugs from the Sea: Proc. 1974, Marine Technology
    Society, Washington, DC, pp. 395-401.

41. Lichter, W., L.L. Wellham, M.M. Sigel and A. Ghaffar, 1979.
    Suppressor activity of splenocytes from mice treated with
    extract of Ecteinascidia turbinata, J. Immunol., 122:8-11.

42. Littlefield, J.W., 1964. Selection of hybrids, from matings
    of fibroblasts in vitro and their presumed recombinants,
    Science, 145:709-710.

43. Maclean J.L., 1979. Indo-Pacific red tides, In D.L. Taylor
    and H.H. Seliger, eds., Toxic Dinoflagellate Blooms,
    Elsevier, North Holland, pp. 173-178.

44. Marderosian, A.D., 1969. Marine pharmaceuticals, J. Pharm.
    Sci., 58:1-33.

45. Martin, D.F. and G.M. Padilla, 1973. Marine pharmocognosy,
    Academic Press, New York.

46. McCumber, L.J., R. Trauger and M.M. Sigel, 1981.
    Modification of the immune system of the American eel,
    Anguilla rostrata, by Ete, International Symposium on Fish
    Biologics: Serodiagnostics and Vaccines, Leetown, WV,
    Develop. Biol. Stand., 49:289-294.

47. Mitchell, R. and L. Young, 1972. The role of
    micro-organisms in marine fouling, Tech. Rept. No. 3, U.S.

Office Naval Res., Contract No. N00014-67-A-0298-0026
NR-306-025.

48. Moore, R.E. and G. Bartolini, 1981. Structure of palytoxin,
    J. Amer. Chem. Soc., 103:2491-2494.

49. Moore, R.E., P. Helfrich and G.M.L. Patterson, 1982. The
    deadly seaweed of Hana', Oceanus, 25:54-63.

50. Moore, R.E. and P.J. Scheuer, 1971. Palytoxin: A new
    marine toxin from a coelenterate, Science, 172:495-498.

51. Moore, R.E., F.X. Woolard and G. Bartolini, 1980. Periodate
    oxidation of N-(p-Bromobenzoyl) palytoxin, J. Amer. Chem.
    Soc., 102:7370-7372.

52. Morse, D.P., H. Duncan, N. Hooker, A. Belour and G. Young,
    1980. GABA-induced behavioral and developmental
    metamorphosis in planktonic molluscan larvae, Fed. Proc.,
    39:3237-3241.

53. Muller, W.A. 1973. Metamorphose-induktion bei
    planulalarven: I. Der bakterielle indukter, Wilh. Roux's
    Arch. Devel. Biol., 173:107-121.

54. Muller, W.E.G., R.K. Zahn, B. Kurelec, C. Lucu, I. Muller
    and G. Uhlenbruch, 1981. Lectin, a possible basis for
    symbiosis between bacteria and sponges, J. Bacteriol,
    145:548-558.

55. Neumann, R. 1979. Bacterial induction of settlement and
    metamorphosis in the planula larvae of Cassiopeia andromeda
    (Cnidaria: Scyphozoa, Rhozostomeae), Marine Ecology, Prog.
    Ser., 1:21-28.

56. Office of Technology Assessment, 1981. Impacts of Applied
    Genetics: Micro-organisms, Plants, and Animals, Report,
    Office of Technology Assessment, U.S. Congress, Washington,
    DC, 331 pp.

57. Old, R.W. and S.B. Primrose, 1980. Principles of Gene
    Manipulation: An Introduction to Genetic Engineering.
    University of California Press, Berkeley.

58. Rains, D.W. and R.C. Valentine, 1980. Genetic Engineering
    of Osmoregulation, Plenum Press, New York.

59. Remmers, E.F., R.R. Colwell and R.A. Goldsby, 1982.
    Production and characterization of monoclonal antibodies to
    cholera toxin, Infect. Immun., In press.

60. Revah-Moiseev, S. and A. Carroad, 1981. Conversion of the enzymatic hydrolysate of shellfish waste chitin to single cell protein, Biotechnol. Bioengineer, 13:1067-1078.

61. Rinehart, K.L., Jr., J.B. Gloer, R.G. Hughes, Jr., H.E. Renis, J.P. McGovern, E.B. Swynenberg, D.A. Stringfellow, S.L. Kuentzel and L.H. Li, 1981. Didemnins: Antiviral and antitumor depsipeptides from a Caribbean tunicate, Science 212:933-935.

62. Ruggieri, G.D, 1976. Drugs from the sea, Science, 194:491-497.

63. Russell, F.E. and P.R. Saunders, eds., 1967. Animal Toxins, Pergamon Press, Oxford.

64. Scharff, M.D. and S. Roberts, 1981. Present status and future prospects for the hybridoma technology, J. Tissue Culture Assoc., 17:1072-1077.

65. Scheuer, P.J., 1973. Chemistry of Marine Natural Products, Academic Press, New York.

66. Scheuer, P.J., W. Takahashi, J. Tsutsumi and T. Yoshida, 1967. Ciguatoxin: Isolation and chemical nature, Science, 155:1267-1268.

67. Schmitz, F.J., Y. Gopichand, D.P. Michaud, R.S. Prasad, S. Remaley, M.B. Hossain, A. Rahman, P.K. Sengupta and D. van der Helm, 1981. Recent developments in research on metabolites from Caribbean marine invertebrates, Pure Appl. Chem., 51:853-865.

68. Schmitz, F.J., K.H. Hollenbeak and D.C. Campbell, 1978. Marine natural products: Halitoxin, toxic complex of several marine sponges of the genus Haliclona, J. Org. Chem., 43:3916-3922.

69. Scofield, V.L., J.M. Schlumpberger, L.A. West and I.L. Weissman, 1982. Protochordate allorecognition is controlled by an MHC-like gene system, Nature, 295:499-502.

70. Scofield, V.L. and I.L. Weissman, 1981. Allorecognition in biological systems, Devel. Comp. Immunol., 5:23-28.

71. Sigel, M.M., 1974. Primitive immunoglobulins and other proteins with binding functions in the shark, Ann. NY Acad. of Sci., 234:198.

72. Sigel, M.M., L.J. McCumber, J.A. Hightower, S.S. Hayasaka, E.M. Huggins, Jr. and J.F. Davis, 1982. Ecteinascidia

turbinata extract activates components of inflammatory
responses throughout the phylogenetic spectrum, Amer. J. of
Zool., In press.

73. Sigel, M.M., L.L. Wellham, W. Lichter, L.E. Dudek, J.L.
Gargus and A.H. Lucas, 1970. Anticellular and antitumor
activity of extracts from tropical marine invertebrates, In
H.W. Youngken, Jr., ed., Food-Drugs from the Sea:
Proceedings, 1969, Marine Technology Society, Washington,
DC, pp. 281-294.

74. Smedes, G.W. and L.E. Hurd, 1981. An empirical test of
community stability: Resistance of a fouling community to a
biological patch-forming disturbance, Ecology, 62:1561-1572.

75. Sochard, M.R., D.F. Wilson, B. Austin and R.R. Colwell,
1979. Bacteria associated with the surface and gut of
marine copepods, Appl. Environ. Microbiol., 37:750-759.

76. Spangenberg, D.B., 1971. Thyroxine-induced metamorphosis in
Aurelia, J. Exp. Zool., 178:183-194.

77. Streisinger, G., C. Walker, N. Dower, D. Knauber and F.
Singer, 1981. Production of clones of homozygous diploid
zebra fish (Brachydanio rerio), Nature, 291:293-296.

78. Sullivan, B. and D.J. Faulkner, 1982. An antimicrobial
sesterterpene from a Palauan sponge, Tetrahedron Letters,
23:907-910.

79. Tachibana, K., 1980. Structural studies on marine toxins,
Ph.D. Thesis, Univ. Hawaii, 157 pp.

80. Tachibana, K., P.J. Scheuer, Y. Tsukitani, H. Kikuchi, D.
Van Engen, J. Clardy, Y. Gopichand and F.J. Schmitz, 1981.
Okadaic acid, a cytotoxic polyether from two marine sponges
of the genus Halichondria, J. Amer. Chem. Soc.,
103:2469-2471.

81. Turlapaty, P., S. Shibata, T.R. Norton and M. Kashiwagi,
1973. A possible mechanism of action of a central stimulant
substance isolated from the sea anemone, Stoichactis kenti,
Eur. J. Pharmacol., 24:310-316.

82. Veitch, F.P. and H. Hidu, 1971. Gregarious setting in the
American oyster Crassostrea virginica Gmelin: I.
Properties of a partially purified "Setting Factor,"
Chesapeake Sci., 12:173-178.

83. Watson, J.D., 1968. The Double Helix: A Personal Account

of the Discovery of the Structure of DNA, Atheneum, New York, 226 pp.

84. Weiner, R. and R.R. Colwell, 1982. Manuscript in preparation.

85. Withers, N.W., 1982. Ciguatera fish poisoning, Ann. Rev. Med., 33:97–111.

86. Woodruff, H.B., 1980. Natural products from microorganisms, Science, 208:1225–1229.

87. Youngken, H.W., Jr. and Y. Shimizu, 1975. In J.P. Riley and G. Skirrow, eds., Chemical Oceanography, Vol. 4, Academic Press, New York, p. 269.

88. Zachary, A. and R.R. Colwell, 1979. Gut-associated microflora of <u>Limnoria tripunctata</u> in marine creosote-treated wood pilings, Nature, 282:716–717.

89. Zachary, A., M.E. Taylor, F.E. Scott and R.R. Colwell, 1979. A method for rapid evaluation of materials for susceptibility to marine biofouling, Int. Biodeterior. Bull., 14:111–118.

SEMINARS

I. AQUACULTURE

Technical and Economic Overview of Aquaculture

Christophe Riboud

Institute de Gestion Internationale
Agro—Alimentaire
Cergy, France

Abstract

Every year large numbers of people die of hunger. Some die
because they do not have a proper diet, many simply lack food.
Most agricultural economists agree that not only do sufficient
food, calories, and protein exist to feed the world's population, but also that the food supply/ demand situation will not
change markedly in the foreseeable future. I argue that price
is at the center of the problem and that people living in
developing countries cannot afford an adequate diet. Aquaculture may contribute to solving this problem. Because it is
expensive to set up an aquaculture enterprise, government
subsidies are needed, but this is only appropriate if aquaculture can offer (1) cheaper protein to developing countries or
(2) a cheaper diet to developed countries. This paper will
address these issues.

Consequences of Increased Meat Consumption

As countries develop throughout the world and the per capita
income increases, people migrate from rural areas to cities. As
urban populations grow, meat consumption increases. This has
been true in the United States and Western Europe and in such
diverse developing countries as Colombia and India. Until very
recently, the Soviet National Academy of Science has consistently refused to increase the meat per capita goal.

While an efficient meat production system may be a reasonable way to produce an economical diet in developed countries,
this may not be so in developing countries. Because urban
workers eat more meat, the cost of feeding them is much higher
than the cost of feeding peasants. In developing countries a
vicious chain reaction results: when people migrate to urban
areas, they tend to eat more meat; not enough meat is produced,

and meat or meat substitutes are imported. The self-reliance of
the country decreases, resulting in the implementation of costly
agricultural policies.

In developed countries, intensive meat production generally
creates dependence upon foreign food imports, which tend to make
up a large part of the national diet. In both developed and
developing countries, the economic consequence of this development is the same: the cost of food rises.

A few years ago, the United States Department of Agriculture
(USDA) attempted to forecast the world food situation (Table I),
assuming that if food prices were to increase, the supply of

Table I. Meat and corn price ratios

		1985	
	Average	I	II
Country and	1973/1974	Status Quo	Free Trade
Price Ratios	1975/1976	Hypothesis	Hypothesis
USA			
Beef/Corn	9.20	13.84	15.10
Hog/Corn	8.68	9.40	10.48
CANADA			
Beef/Corn	10.47	13.92	15.16
Hog/Corn	12.28	12.81	15.47
OCEANIA			
Beef/Corn	7.98	18.77	19.41
Hog/Corn	13.70	24.90	23.39
EEC[1]			
Beef/Corn	9.48	9.24	9.65
Hog/Corn	8.32	9.10	9.70
EEC[2]			
Beef/Corn	15.03	15.01	15.74
Hog/Corn	16.15	10.95	11.73

Source: USDA, 1978. Alternative Future for World Food in
1985, Foreign Agricultural Economic Report.No. 146, Note
5, Washington, DC.

[1] European Economic Community: Belgium, France, West
Germany, Italy, Luxembourg, Netherlands
[2] Same as EEC[1] plus Denmark, Ireland, Great Britain

food would also tend to increase. The USDA concluded that two
commodities which might be in short supply and therefore might
cause prices to rise rapidly are beef and pork. Two hypotheses
used by the USDA - "Status Quo" (no change in agricultural
policies) and "Free Trade" - led to the prediction that food
exports and imports will increase throughout the world. In
fact, it is predicted that, except in the European Economic
Community, meat prices will increase faster than grain prices.

The Role of Fish Consumption

To alleviate the food price situation, great expectations
have been placed in intensive fish production. Arguments
support the claim that fish should be used more extensively as a
meat substitute. Raw protein from fish is considerably cheaper
to raise than the same amount of protein from pork (Table II).
Less energy is required to produce one gram of fish protein than
meat protein (Table III). Moreover, fish products are rich in
vitamins and low in sodium and fats. Fish protein is a high
quality protein; the only proteins used more efficiently by
humans are those from eggs and milk (Table IV).

If fish has so many advantages, why isn't more used? Table
V illustrates a paradox: almost two-thirds of the population in
developing countries, where people are underfed, receive more

Table II. Protein cost by type of protein

	$/Kg
Fish	0.72
Coarse Grains	1.21
Rice	1.52
Shellfish	2.10
Wheat	2.15
Poultry	4.27
Beef	4.55
Hogs	5.06

Source: Bell, Canterbury, 1976. Aquaculture for the Developing Countries,
Ballinger Publishing Co.

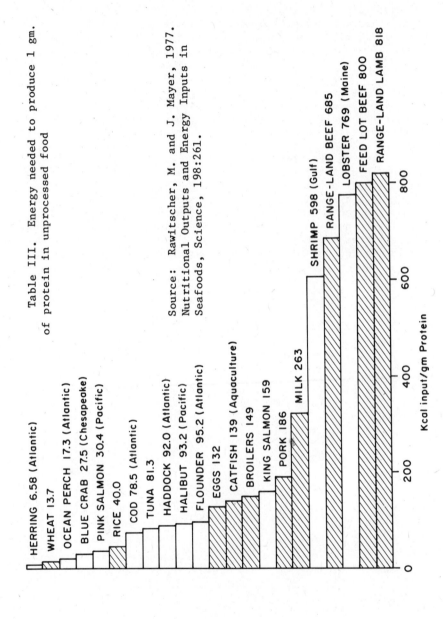

Table III. Energy needed to produce 1 gm. of protein in unprocessed food

Source: Rawitscher, M. and J. Mayer, 1977. Nutritional Outputs and Energy Inputs in Seafoods, Science, 198:261.

HERRING 6.58 (Atlantic)
WHEAT 13.7
OCEAN PERCH 17.3 (Atlantic)
BLUE CRAB 27.5 (Chesapeake)
PINK SALMON 30.4 (Pacific)
RICE 40.0
COD 78.5 (Atlantic)
TUNA 81.3
HADDOCK 92.0 (Atlantic)
HALIBUT 93.2 (Pacific)
FLOUNDER 95.2 (Atlantic)
EGGS 132
CATFISH 139 (Aquaculture)
BROILERS 149
KING SALMON 159
PORK 186
MILK 263
SHRIMP 598 (Gulf)
RANGE-LAND BEEF 685
LOBSTER 769 (Maine)
FEED LOT BEEF 800
RANGE-LAND LAMB 818

Kcal input/gm Protein

0   200   400   600   800

Table IV.   Protein producing food

	% Net Utilization of Proteins by Human Beings	Efficiency Ratio on a Protein Basis
Meat	24	-
Hamburgers	42.7	1.65
Wheat	40.3	1.43
Yeast	55.6	2.24
Soja	61.4	2.32
Rice	57.2	2.18
Milk	81.6	3.09
Beef	66.9	3.2
Eggs	93.5	3.9
Fish	79.5	3.55
Fish meal	65.8	3.42

Source: Whitaker and Tannenbaum, 1977.   Food Proteins, AVI Publishing Co.

than 40 percent of their protein from fish products. More than
three-quarters of the population in developed countries, where
people suffer from unbalanced diets, get less than 30 percent of
their protein from fish. An enormous proportion of the population in centrally planned economies receives a significant
amount of its protein from fish products; yet in its most recent
plan, the Russian government has finally decided to increase
meat production. It would seem that economic development is far
from being associated with increased fish consumption!
    Fish as a source of protein would help solve the needs of
developing countries. People in these countries consume only 38
percent of the world fish catch. The percentage of calories
obtained from meat and fish is approximately four times smaller
in developing than in developed countries.

Table V. Contribution of fish to the world consumption of proteins (1972-1974)

Fish Contribution	% of Population in			
	Developed Countries	Developing Countries	Centrally Planned Economies	World
40% +	16	57	5	33
30-40%	6	2	86	30
20-30%	1	9	3	5
10-20%	38	22	3	19
0-10%	39	10	3	13

Source: Krone, W., 1979. Fish as Food, Food Policy, 4:259.

    Another problem with fish production is the economic barrier that makes entry into the fishing industry extremely difficult. Although expensive meat products make the development of a fishing industry an attractive proposition in developed countries with long coastlines, high energy costs and enormous investments in vessels require a large capital commitment. In recent years the number of companies harvesting fish has declined and more of the fish catch is accounted for by giant firms, whether Japanese, American, or one giant firm called the Soviet Union. Of the 1,144 large fishing vessels (greater than 2,000 tons, GRT's) in the world, Japan and the Soviet Union own 869. Of the 131 large factory ships in the world, Japan and the Soviet Union own 117. The fishing industry is indeed very capital-intensive and concentrated, and this makes it difficult for developing countries to participate in this industry on a significant scale.
    Worldwide, the annual fish production is roughly 70 million tons (compared to 120 million tons for meat) and this amount slightly exceeds demand. By the year 2000, some experts forecast that demand for fish products may increase at an annual rate of 3.5 percent, while catch and fish processing capacities will only increase at an annual rate of 1 percent. This analysis implies that fish will be in short supply, and that its cost should rapidly increase. In France, fish is now more expensive than meat, and its price tends to increase more

Table VI. French CPI[1] (1978–1979) for different foods

	1978	1979	Price Change 1978/1979
Beef	215	229	+6.5%
Veal	194	206	+6.2%
Lamb	210	225	+7.1%
Pork	187	201	+7.5%
Poultry	196	210	+7.1%
Fish	240	266	+10.8%

Source: Comite Central des Peches Mari-
    times, 1980. Rapport du Groupe de Travail,
    "Mer et Littoral," Preparation du 8 ieme
    plan, La Documentation Française, p. 143.

[1] Consumer Price Index

rapidly than meat prices (Table VI). A French study has shown
that fish is associated with high income elasticity and even
higher price elasticity (La Consommation Alimentaire en 1977,
Les Collections de l'Insee, Serie Menage). The average income
elasticity for fish is estimated to be 0.8, meaning that when
real income increases 5 percent, demand for fish products
increases 4 percent. Price elasticity for fish is estimated to
be 1. Therefore, a 5 percent increase in real income coupled
with a 7 percent increase in fish prices would result in a 3
percent decrease in fish consumption.

The situation is further complicated by the irregular nature
of the fish catch throughout the year. Price variations are
also much larger than income variations. Consumers are always
more sensitive to an increase in food prices than to the fact
that their income may have recently increased. In the United
States, the picture is slightly brighter. Table VII shows
parallel increases in meat prices and fish consumption, which
has grown more rapidly than consumption of most foods except
chicken. Between 1967 and 1979, for instance, the index of per

Table VII. Prices and consumption of fish products in the U.S.

| | Unit Value for Fish Products | | | | | CPI | CPI Food |
| | Canned | | Fresh and Frozen Packaged | | | | |
	$/1b.	100=1967	$/1b.	100=1967		1967=100	1967=100
1967	.44	100	.42	100		100	100
1970	.55	125	.56	133		116	115
1975	.77	175	1.05	250		161	175
1979	1.22	277	1.39	331		217	235

	Per Capita Consumption 100=1967[1]				Per Capita Consumption In retail weight equivalent		
	Food	Meat	Poultry	Fish	Meat	Poultry	Fish
1967	100	100	100	100	158.5	45.5	14.2
1970	102	104	107	111	164.8	48.9	15.8
1975	101	101	108	115	158.4	49.3	16.6
1979	105	100	129	124	159.5	59.0	17.6

Source: USDA, 1980. Agricultural Statistics, 1980, U.S. Government Printing Office, Washington, DC.

1 Consumer Price Index

capita U.S. food consumption increased from 100 to 105, meat
consumption remained at 100, poultry increased from 100 to 129,
and fish from 100 to 124. This is a significant achievement for
fish products, particularly when price changes are taken into
account. The consumer price index (CPI) for food increased from
100 to 235 during these years, while the price index for canned
fish products increased from 100 to 277, and for fresh and
frozen fish from 100 to 331. Despite the fact that fish prices
are increasing faster than food prices in general, Americans eat
more fish.

Table VIII shows, however, that half of the fish sold in the
United States is imported. Two thirds of fish consumption
occurs outside the home. I would contend that price elasticity
for fish products is higher for restaurants and institutions
than for individual consumers. If meat prices, and particularly
prices for white meat, continue to decline as they have done
recently and are projected to do, fish consumption may decline
in the next few years.

Table VIII. 1978 U.S. sales of fishery products
(billions of $)

	National Products	Imported Products	Total	% of Total
Retail	1.6	2.0	3.6	31.8
Restaurants	4.1	3.1	7.2	63.1
Institutional	0.3	0.3	0.6	5.1
Total	6.0	5.4	11.4	100
% of Total	52.4	47.6	100	
Consumption per Capita in Value ($)			52.50	
In Volume (lb.)			13.4	
Population (M)			218	
Exports (M$)			831.7	

Source: Marine Fisheries Review, December, 1979, p. 2.

The Role of Aquaculture

The difficulties mentioned above make it very interesting in developing and developed countries to investigate alternative ways of producing fish protein. Aquaculture can be economically justified only if the initial investment to enter the fishing industry is lowered, and the cost of fish products to the consumer is reduced.

In 1975, aquaculture accounted for less than 10 percent of world fish production. Almost four million tons of fish were produced by aquaculture, 90-92 percent in fresh water. Fish (primarily freshwater varieties) make up about two thirds of aquaculture products. Japan, China, and India are the greatest practitioners of this technique. Organization for Economic Cooperation and Development (OECD) experts recently estimated that the total production of fish by aquaculture was about nine million tons in 1979.

The Food and Agriculture Organization of the United Nations is optimistic about aquaculture's future because of the organization's high estimates of unused resources (Table X). Unfortunately, some of these resources are quite difficult to handle. Krill, primarily consumed in Japan and the Soviet Union, accounts for about 100 million tons of unused resources, but must be processed without delay, i.e. at the place of harvest, because of the harvest's rapid deterioration; exploitation of krill also requires long transportation from the fishing grounds in the Antarctic.

Today, most aquaculture is limited to very simple technologies - growing fish with low yields. These technologies require low investment, yield a product with low commercial value, and are very labor-intensive. It is hard to present this type of aquaculture as the future of fish farming.

The economic viability of aquaculture depends on technical knowledge about various fish species. Technical knowledge requires an investment in research. Japan, the United States, Norway, France, the Netherlands, and other European countries believe in the industrial possibilities of aquaculture and have made large research investments. These efforts have led to various degrees of knowledge, depending on the fish species. From the French perspective, the main categories of fish that are useful for aquaculture are rainbow trout, sea bass, coho salmon, Dover sole, eel, bream, mullet, turbot, oysters, and crustaceans. For each of these species, there are problems and there are hopes. For example, in France, the raising of rainbow trout has been by far the most successful operation. However, a serious problem in growing trout is the fish's poor tolerance to varying water temperatures.

Table XI summarizes the important technical functions and the implications that must be considered. If the physiology of reproduction is improved, the survival rate at the larval stage will increase. Improving the nutritional quality of feed increases conversion and survival rates. Many problems remain

Table IX. Aquaculture production in the world (1975)

Group	Tons	%	Characteristics	Three Main Producers in %	France
Algae	1,055,000	17.5	Salt water	Japan........ 48 China........ 28 Korea........ 23 99	–
Shells	987,000	16.2	Salt water	Japan........ 30 Spain........ 16 USA.......... 13 59	88500 T, 5th rank
Crustaceans	16,000	0.3	Salt water	India........ 25 Indonesia.... 25 Thailand..... 21 71	–
Fish	3,980,000	66	90 to 92% in fresh water	China........ 55 India........ 12 USSR......... 5 72	15000 T,
Total	6,029,000	100	40% in fresh water 60% in salt water	China........ 41 Japan........ 16 India........ 8 65	103500 T, 12th rank

Source: Rapport sur L'Aquaculture, Rapport au Conseil Economique et Social, Session de 1981, Seances du 24 et 25 Novembre, 1981, Publie 1982, p. 172.

Table X. "Potential" fishery resources as given by FAO[1] (in million tons)

Resource	Used	Unused	Total
Fresh water	10.2	5	15
Salt water Conventional	57.4	52.7	120.6
"Unconventional"	1.3	250	236
Total	68.9	307.7	371.6

Source: same as Table IX.

[1] Food and Agricultural Organization

Table XI. Functions and performance measures in aquaculture

Functions	Performance Measures
Physiology of reproduction	− Survival rate at larval stage
Physiology of feedstuff	− Conversion rate − Survival rate
Genetics	− Survival rate (resistance to diseases) − Conversion rate − Growth rate
Pathology	− Survival rate
Ecology	− Survival rate − Growth rate
Technology	− Growth rate − Survival rate − Feedstuff loss and conversion rate

in growing different species such as sole, bream, eel, and
salmon either at the reproductive or larval stage. These
species have low fry survival rates. Additional problems exist
in the growth stage, where insufficient knowledge of the
pathology, physiology, nutrition, and genetics of these fish
makes for low conversion rates and uncertain survival rates.

Not being a specialist in the scientific and technological
aspects of aquaculture, let me now address these important
questions: Does aquaculture hold good economic prospects? Are
the survival and conversion rates for the different species of
fish, as they stand today, good enough to make an aquaculture
farm a viable commercial proposition?

There are two points of view from which to answer this
question. One is that of a government, clearly a different
point of view than that of a private firm. Despite the creation
of exclusive economic zones, the United States still runs an
impressive deficit in fishery products. Table XII shows that
between 1975 and 1979 the deficit in fishery products doubled
for the United States, increasing from about $1 billion to
almost $3 billion. Fish could be considered a strategic
commodity.

Table XIII shows that France is in a very similar
situation. Our net trade deficit has tripled, despite the fact
that we have the longest coast in Europe. It is no secret, at
least in France, that our fishing industry is in very bad shape.

Table XII. Net trade balance for fishery
products, US, 1975/1979 (in millions of $)

	1975	1979
Edible Products	1000	1647
of which:		
Fresh and salt water fish,	497	1373
Shellfish, fresh or frozen	502	817
Canned fish and shellfish	91	118
Non Edible Products	233	1081
Total	1323	3389

Source: USDA, 1980. Agricultural Statistics,
1980. U.S. Government Printing Office,
Washington, DC.

Table XIII. Fishery products in France (in million FF)

	1970		1975		1978	
	Production	Net Trade Balance	Production	Net Trade Balance	Production	Net Trade Balance
Salt water fish, fresh and frozen	925.0	-253.2	1,440.1	-439.4	2,229.5	-713.7
Smoked and cured products	71.8	11.5	22.7	-29.9	11.9	-132.0
Crustaceans	148.5	-111.4	266.8	-245.5	328.2	-487.3
Shellfish	42.6	-59.5	878.7	-166.0	950.3	-193
Algae	6.1	12.3	7.1	18.5	6.9	-
Oil and meal	2.7	-101.0	4.0	-98.3	6.2	-102.0
Total saltwater	1,196.7	-501.3	2,619.4	-960.6	3,553.0	1,627.94
Freshwater fish				-204.8		-348.35
Canned fish		-275.7		-418.2		-732.05

Source: Same as for Table VI, pp. 132-134.

55

The most interesting situation exists in Japan. Before
1978, this country accounted for more than 15 percent of world
fish production. After the exclusive economic zones were
declared in 1978, the value of Japanese imports increased 28
percent in one year. Between 1977 and 1978, the Japanese trade
deficit jumped from $1.6 billion to $2.3 billion.

The stakes are extremely high for fishing countries. It is
not surprising that very important sums of money are spent by
governments for the development of aquaculture. Japanese aid to
aquaculture is extremely impressive (Table XIV) with a 40
percent increase between 1978 and 1979. As far as I know, this
commitment has not decreased.

It is difficult to understand whether all of these efforts
make real economic sense, but I would like to point to one
possible positive outcome. In France, sole is caught in the
traditional way, harvested from boats, whereas 100 percent of
French trout is produced by aquaculture. While the prices of
both were about the same in 1969, sole's traditionally-harvested
price has increased much faster than the freshwater trout's.

Table XIV. Government funding of aquaculture in some
countries (in millions of $)

U.S.	1979	1980	1981
Federal	3.7	4.3	3.7
Matching	3.3	3.0	2.7

Source: Price, K., Aquaculture, unpublished manuscript.

Japan	1977 1978	1978 1979	Increase
Aid to aquaculture	6.8	10	47%
Aid for coastal fishing	20	34	70%

Source: Same as Table IX, p. 176.

Also, the price of trout has been much more stable. Thus,
aquaculture can result in lower and more stable prices. All of
this points to obvious advantages for national programs.

The economics of aquaculture can also be examined from the
point of view of individual companies. Table XV gives the
results of a study by the Commission of the European Economic
Community of the production costs for fish raised on an
experimental farm. Salmon is the species closest to commerical
operation; feed was found to be the largest component of the
production cost.

To confirm these figures, I conducted a small survey of six
aquaculture firms in France. The percentages of the various
items in the total costs appear in Table XVI. These firms have
been in operation, on the average, for four to five years so
they are past the development stage. The table shows the
critical importance of two items: fry and feed. It also shows
that, as of today, labor remains important, the reason being
that the development of aquaculture farms is just beginning, and
economies of scale have not yet been established.

All of the French firms surveyed shared the common feature
that they could not have stayed in operation without government
subsidies. Oyster production is the only French aquaculture
operation economically viable today, but it is not included in
the sample surveyed.

Table XV. Cost of production in marine farm (%)

		Sole	Turbot	Salmon
1.	Financial Expenses	31	35	15
2.	Food	20	14	40
3.	Fry	15	7	14
4.	Personnel	11	13	20
5.	General Expenses	9	12	5
6.	Energy	4	7	1
7.	Various	9	12	5
		100%	100%	100%

Table XVI.   Percentage of various cost items

	Mean Cost	Cost Range
Fry	25	20–31
Food	24	18–31
Labor	20	17–25
Operating costs	14	11–18
Commercialization	5	3– 7
Management	2	2
Taxes	–	–
Financial expenses	3	3
Amortization	7	3–15

Finally, I tried to evaluate how sensitive a firm could be to uncertainty in technical parameters such as survival and conversion rates.   Table XVII shows a theoretical profit and loss statement for a 30-ton-per-year sea bass operation in France. A 20 percent profit margin is probably not realistic, otherwise people would get into this business. This analysis shows the importance of the expenses of initial fish stock. Add to that the purchase of fry, representing 60 percent of the operating expenses. Adding the cost of feed accounts for 71 percent of operating expenses. The survival rate (a survival rate of 70 percent for sea bass was assumed) could go down because of an increase in the water temperature or the introduction of a disease. If survival declined to 50 percent, the operation would be destroyed in one year. These operations, therefore, are extremely sensitive to the value of the initial stock.

In conclusion, aquaculture investment costs are still extremely high, and additional research must be conducted to ensure that estimated expenses are realistic and will not double if something goes wrong.

A second point which must be addressed is: Can aquaculture
produce the kind of fish the consumer demands? The American
example indicates that the highest consumption of fish is eaten
outside the home in a form that is easy to prepare by some
industrial process. This is not the case with trout. In
France, where the aquaculture of trout has been quite successful, an oversupply now exists. It seems to me rather important
to start working on fish species that industry feels comfortable
in transforming, processing, and distributing to the consumer.

Table XVII. Sea bass operation

	Expenses
Initial stock	52
Purchase of fry	8
Food	11
Personnel	12
Energy	1
Other operating expenses	6
Financial expenses	10
Amortization	10
	100
Profit	20

Table XVIII. Critical variables for some fish

	Trout (sea)	Bass	Coho Salmon
Breeding length	10-18 months	2-3 years	
Average sale size	200 g 400 g 1 kg	300 g	300g 1.5kg
Cost of fry	.25 ff	1-2 ff	14.25 ff at 80
Cost of food	4-4.5 ff/kg	4 ff/kg	4-4.5 ff/kg
Conversion rate	1.8-2.2	1.5-2.9	1.8 until 500 2.2 until 1 kg

Survival rate
  Larvae
  Growth
	
	95%
	(down to 50%
	during summer)
		
		40-50%
		70%	

Progress and Problems with Recirculating System Aquaculture

William S. Gaither

University of Delaware
Newark, Delaware

Abstract

Aquaculture research at the University of Delaware has
focused on recirculating systems for bivalves. Both algal and
artificial diets for bivalves have been examined. There are
differences in digestibility of algal food species attributed
mainly to the ability of the bivalve to break down the algal
cell wall. It has also been found that a mixture of algal
species generally results in more rapid growth of bivalves than
when a single species of algae is used. Microencapsulation
techniques have been developed for artificial foods.
Several generations of recirculating systems have been
designed and tested with the result that oysters can be spawned
and grown as "singles" up to approximately 15mm diameter in
approximately six to eight weeks after set.

I have taken this occasion, as a research administrator and
to some extent a research policymaker, to describe the progress
we have made at the University of Delaware in recirculating
system aquaculture for shellfish, and also to look into the
future and suggest potential applications of the acquired
knowledge as well as areas requiring further research.

Historical Perspective

This paper will be a brief history of aquaculture research
at the University of Delaware. In 1966 the National Sea Grant
College and Program Act became a federal statute, and in 1968
Delaware began its Sea Grant-funded research to improve the

oyster industry in the Delaware Bay, which had been ravaged by
the MSX disease (Minchinia nelsoni) (Haskin, Stauber, and Nackin
1966) and adversely affected by at least a century of water
quality degradation in the estuary.

Our approach to this problem was influenced by the report
published in June 1968 by the American Cyanamid Company titled
New Engineering Approaches for the Production of Connecticut
Oysters. We planned our initial Sea Grant research in the
1968-1970 period to learn as much as we could about recirculating systems for growing shellfish. This approach required that
biologists join forces with engineers to build and operate a
recirculating system for the controlled, intensive culture of
oysters. Engineers were also involved in systems studies to
help guide the research as quickly as possible from the laboratory to commercial production of various bivalve species.
I was interested to re-read a statement made by Dr. Donald
Maurer, a biologist and a founding member of the research team,
in our April 1970 Sea Grant proposal. He wrote:

> One important point, however, which is the singular thrust
> of the University's Sea Grant Program, is the concentrated,
> inter-disciplinary effort towards developing a closed
> system. Few organizations have approached shellfish
> production, or any other problem from an integrated, multi-
> disciplinary point of view, and even fewer people are con-
> cerned with a closed system.----It may not be possible to
> design systems to commercially produce oysters in a con-
> trolled environment in a few years, but the real value of
> the work lies in the future.

That was a fair statement in 1970 and, in spite of substantial progress during the intervening years, it is a reasonable
statement to repeat in March 1982. Rather than carry you
through a chronological recitation of the successes and failures
during nearly 14 years of research, I will address a few
selected topics that are relevant to why and how we attacked
this problem and where we see shellfish aquaculture developing
in the future.

Why Recirculating System Mariculture?
The first step was to define the problem, and the most
promising approach to solving the problem, as clearly as
possible. The problem was the uncertainty of culturing
shellfish successfully in the uncontrolled natural environment,
which is subject to the vagaries of nature as well as to the
carelessness of humankind. The approach chosen to solve the
problem was to emulate argicultural production systems where
domesticated plants and animals, bred to possess desirable
characteristics, are provided with optimum food and environmental conditions consistent with the price consumers are
willing to pay for the final product.

The second question which needed to be answered was why should the University of Delaware undertake this line of research? In answering that question we typically apply a three point test.
1. Is the problem or opportunity one of substantial significance to the nation and humankind?
2. Is the problem one to which others are as yet giving inadequate attention?
3. Is the problem one for which the University of Delaware has a unique advantage--or at the very least no significant disadvantage--in addressing?
Our response to each of these questions was positive.

Can Bivalves be Reared in Captivity?

To begin the research program we first had to answer this question. In a 1969 Sea Grant conference organized by Drs. Price and Maurer (1971) addressing the issue of the Artificial Propagation of Commercially Valuable Shellfish, it became clear that a great deal was already known about holding, conditioning, and spawning adult oysters, as well as culturing the larvae through settlement to metamorphosis, into juvenile stages. What was not known, or at least demonstrated, was how to put all of the knowledge together into a dependable operating system from which a specified output rate and quality could be guaranteed.

In spite of the large amount of information available on the general biology of bivalves, the literature was unable to provide fundamental information necessary to specify (1) the rate of consumption of food, oxygen, and dissolved chemicals by bivalves, (2) the rate of production of wastes by the animals, and (3) the tolerance of the animals to various water quality conditions, particularly those resulting from accumulation of their own waste products.

Our first step in quantifying information necessary for engineering design was to develop mathematical models relating bivalve intakes and outputs. From these relationships, system design specifications were developed (Epifanio, Srna, and Pruder 1975). In addition, we conducted extensive experiments on the tolerance of bivalves to their own waste products and were able to define safe levels of compounds such as ammonia, nitrites, nitrates, and orthophosphates (Epifanio and Srna 1975).

By 1973 our researchers were able to spawn and grow both oysters and clams in the laboratory. Annual growth rates that were equal to or greater than those found in nature were achieved (Epifanio, Logan, and Turk 1975). By 1975 we had successfully grown clams and oysters to market size in a recirculating system, although the number of animals reared in this way were few, due to difficulties in providing a dependable source of nutritious food. In a few cases growth rates were spectacular, and we were optimistic that progress would be rapid. By 1978 some of our brood stock comprised animals which

had been spawned and cultured in captivity for several
generations.

Diet for Bivalves

From the outset of our research we knew that bivalves would
grow if fed an algal diet. What was not known were bivalve
requirements for (1) quality, (2) quantity, and (3) supplements. In our research program the problem of bivalve nutrition
has been a most difficult and perplexing area. The primary
component of the problem has been understanding the nutritional
requirements of bivalves. The other component of the problem is
providing enough algal food for the animals in a recirculating
system.

The nutrition question has been addressed two ways. The
first has been to determine which algal diets are most suitable
for bivalves and to try to understand the underlying reasons for
differences in algal food quality. The second approach has been
to create a complete diet for defined dietary components so that
classical deletion experiments may be conducted through manipulation and successive withdrawals of components of the diet so
as to examine the nutritional requirements of the animals.

Since algal diets were examined initially, and algae is the
food produced in our most sophisticated recirculating system, I
will summarize that line of research first. Fortunately for the
Delaware team, much work on algal diets for larvae had been
undertaken in other laboratories and research stations. The
principal emphasis of the Delaware algal research has been on
the food value of various species and combinations of species
for bivalves. Differences in the digestibility of algal food
species have been found. These have been attributed mainly to
the ability of the bivalve to break down the algal cell wall. A
second important finding of this research has been that a
mixture of two or more algal species generally results in more
rapid bivalve growth than when either species is fed alone
(Epifanio, in press).

The second part of our algal research focused on ration
size. In other words, how much food does an animal require to
achieve the maximum efficiency of growth? Several approaches to
this problem were undertaken, including investigation of cell
density effects, continuous versus discontinuous feeding, and
absolute requirements for food in a fixed time period (Epifanio
and Ewart 1977; Romberger and Epifanio 1981). The outcome of
this research, together with other results, has been an empirical equation that predicts the maximum daily algae requirement
of oysters of any particular size (Pruder, Bolton, and Faunce
1977).

Another component of our nutrition research was to correlate
food ingestion and growth with temperature (Muller 1978). One
of the most interesting findings of this research was that
growth was highly dependent on both ration and temperature.
Under high temperature (28°C) and low ration conditions, shell

growth was enhanced while meat growth was retarded; whereas with a high ration, at the same high temperature, the ratio of meat to shell was enhanced.

At the low end of the temperature range tested (18°C) a low ration resulted in modest growth and the highest ration produced only a slightly greater growth rate than with the lowest ration. This finding is obviously important to the operation of a recirculating system where it is necessary to maximize food utilization efficiency by selecting the most appropriate temperature and ration size.

The potential use of non-algal diets was examined. These studies included various formulated feeds, several types of starch, ground whole cereals, and <u>Torulopsis</u> yeast. The yeast proved to be the only promising non-algal food tested (Epifanio 1979).

Artificial Diets

In addition to the algae studies a more fundamental approach has been taken to the question of bivalve nutrition. The research program consists of two complementary efforts, the first now largely completed, and the second just underway (Langdon 1981). The objective of the program is to design and test artificial diets for bivalves so that all the essential constituents and the optimum dietary levels of these constituents can be determined.

To succeed in this approach the first requirement was to obtain a dependable supply of axenic oyster larvae. It was important that these larvae be available for nutrition experiments without the presence of any other living organisms which may have acted as an undefined food source and without the use of antibiotics which may have altered the oysters' nutritional requirements. This has been accomplished (Langdon 1981).

The second step, now well underway, is to develop artificial diets that will support oyster growth. Several microencapsulation techniques have been developed and tested with satisfactory results. Dr. Langdon (1981) observed:

It is now technically possible to encapsulate all the potentially important constituents of artificial diets, namely, high molecular weight proteins and starches within nylon-protein walled capsules; lipids and lipid soluble components within gelatin-acacia walled capsules and low molecular weight water soluble components such as amino acids, minerals and vitamins within lipid walled capsules. Such capsules are especially useful for nutrition studies with filter-feeders since particle breakdown and nutrient leaching are major problems with micro-sized particles. Instability and leaching are also problems with feeds for large particle feeders such as shrimp and lobsters and encapsulation techniques may prove equally useful in this area.

It is our hope and expectation that this line of research
will permit us to pursue a more fundamental and rewarding line
of bivalve nutrition research. This is the kernel of the
problem now. The question is, however, whether microencapsulation diets will be economically feasible in the long term.
The most promising application at the moment would be their use
in providing the oysters with supplements to the algal diets.

Recent experiments indicate silt particles play an important
role in sustained rapid growth of oysters (Ali 1981; Pruder and
Ewart, in press).

Configurations for a Recirculating System

Up to this point I have addressed only the essential
husbandry questions which must be satisfactorily answered if a
recirculating aquaculture system for growing bivalves is to be
successfully constructed and operated. Let me review. First,
can bivalves be reared in captivity? The answer is yes.
Second, what should bivalves be fed? The answer is algae will
do, but there are probably better and cheaper non-algal feeds
which we have not yet identified.

The question of microbial diseases was studied and found not
to be a significant problem in the areas investigated, though it
may be in the future when commercial systems are operated at
even higher densities and for long periods of continuous
production. For example, it is possible that bacteria play a
role in the instability of larger algae cultures.

Also, the question of accelerated growth through species
selection and hybridization was recognized as a place where
refinements could be made, but it was not considered central to
the question of building and operating a recirculating
aquaculture system for growing bivalves.

The first efforts to construct a laboratory scale system
employed aquaria for the shellfish, and flasks and later,
carboys for growing algae. After evaluating these early crude
experiments conducted in the late 1960s and early 1970s we set
out to develop better system components before putting together
the next generation recirculating system.

During the period between 1972 and 1975 three successive
small scale laboratory systems were constructed and operated.
The configuration of the first system is shown in Figure 1
(Pruder 1976). The primary purpose of this system was to
operate and understand the function of a biological filter in a
recirculating system. The growing tank contained 6000 liters of
water and had a false bottom on which rested a sand filter.
Water from the oyster growing tanks containing animal waste and
excess food were pumped through a protein skimmer and then
alternatively treated with (1) activated carbon, and (2)
ultraviolet light. Ozone treatment was considered but not
used. Algae for food was grown in a separate system using
artificial light and then transferred in batches to the growing
tanks. The combined growing tank/biological filter was found to

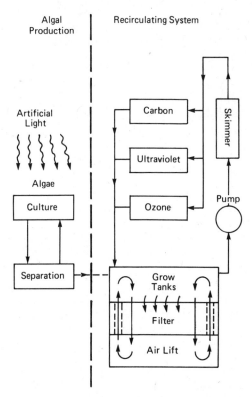

Figure 1:   Configuration I

be unsuitable for the mariculture of filter feeders. Attempts
to harvest algal cells intact from the algal medium efficiently
and reproducibly were unsuccessful. The protein skimmer, while
apparently useful, was not adequately evaluated in this system.

The second generation system shown in Figure 2 was designed
to conduct simultaneously a series of parallel nutritional
studies on both oysters and clams. It featured separation of
the filter and growing tanks. Algae were still grown in a
separate subsystem using artificial light and batch fed to the
oysters. The principal use of this system was for comparative
studies of algal nutritive value. Ozone treatment of
recirculating water was omitted, but activated carbon and
ultraviolet light were retained. The waste treatment systems
were operated on an intermittent basis controlled by a timer.

The third configuration is shown schematically in Figure 3
(Pruder, Bolton, and Faunce 1977; Pruder 1976). It was designed
to support a production rate of approximately four bushels of
bivalves per year. This system was the first designed to
provide algae continuously to the oysters in amounts sufficient
to meet their nutritional needs. In this system the inputs to
the algae and cultures included artificial light, CO_2,

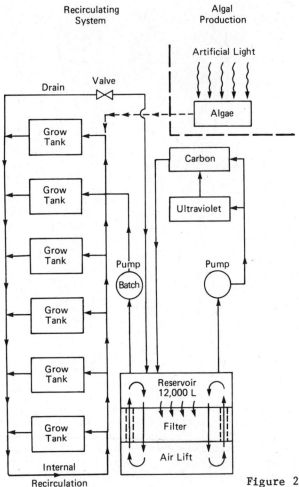

Figure 2:   Configuration II

calcium, and micronutrients (Sick and Johnson 1977; Sick 1981).
Make-up water was added as needed when settled wastes from the
oyster tanks were removed.   The use of artificial lights was
found to be too costly for scale-up and the system was
redesigned to use sunlight.

   The matter of shell development in oysters growing in
recirculating systems was also addressed, but since shells from
recirculating systems were, in all regards except size, normal
in comparison with shells of naturally grown oysters, and there
were no obvious deformities or anomalies associated with
culturing, the matter was not pursued further (Palmer and
Carriker 1979; Carriker and Palmer 1979; Carriker, Palmer, and
Prezant 1980).

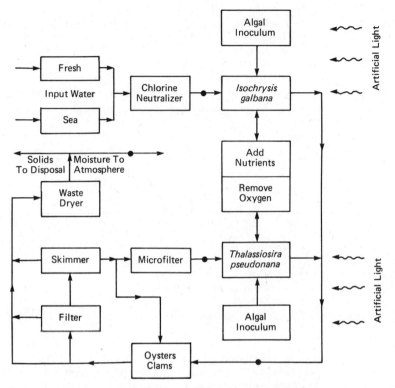

Figure 3: Configuration III

The tastes of both oysters and clams produced in a recirculating system were tested. Both oysters Crassostrea virginica and clams Mercenaria mercenaria harvested from the controlled environment system were organoleptically compared to their wild counterparts. Sensory evaluations considered appearance, texture, flavor, aroma and overall acceptance. In the case of oysters, the greenhouse samples were preferred in all categories over their wild counterparts. In the case of clams, there were no significant differences found between greenhouse and wild animals although there was a slight preference for the flavor of clams produced in the natural environment (Thoroughgood 1979; Hicks 1981).

The fourth generation (or greenhouse) system was the so-called "50 bushel system" since it was designed to support the annual production of that quantity of oysters per year fed on an algal diet. This was put into operation during the spring of 1976 (Pruder, Bolton, and Faunce 1977). Due to the large quantities of algae required, the system was designed to be housed in a dual-skinned vinyl covered greenhouse shown in

William S. Gaither

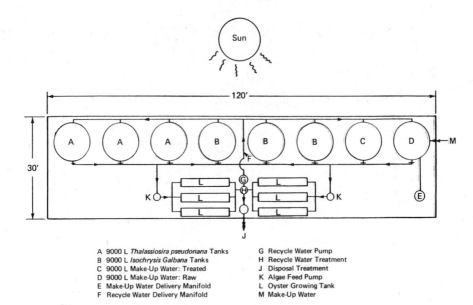

A 9000 L *Thalassiosira pseudonana* Tanks   G Recycle Water Pump
B 9000 L *Isochrysis Galbana* Tanks   H Recycle Water Treatment
C 9000 L Make-Up Water: Treated   J Disposal Treatment
D 9000 L Make-Up Water: Raw   K Algae Feed Pump
E Make-Up Water Delivery Manifold   L Oyster Growing Tank
F Recycle Water Delivery Manifold   M Make-Up Water

Figure 4:   First Generation Configuration IV Greenhouse

Figure 4.   Algae were grown in 3.6-meter-diameter by 1-meter-
deep pools. Two species of algae, Thalassiosira pseudonana (3H)
and Isochrysis galbana, were grown in separate sets of pools.
These were the first cultures of this size (9000 liters each)
and the first to rely on sunlight instead of artificial light.
In this environment, with normal day-night cycles, new problems
developed with frequent "crashes" or death of the majority of
algae in the pools. These problems led to the study of the
biochemistry of photorespiration and the development of equipment to monitor and control automatically CO_2, O_2 and pH
(Pruder 1978). The objective was to control photosynthesis and
photorespiration to (1) prevent "crashes" of the large cultures
and (2) to prevent the loss of fixed carbon in the algae. Algae
were fed directly from the growing pools into the oyster growing
tanks. After water passed out of the growing tanks it contained
dissolved oyster wastes and some algae not removed by the
animals in the growing tanks. These waste materials were
removed by a protein skimmer, and the water was recirculated
directly to the algae growing pools.
     The fifth generation system is shown in Figure 5. This
system was designed by a chemical process engineer using design
data developed from the previous systems. Two species of algae
were grown simultaneously in two sets of five tube reactor
modules. The details of the modules are shown in Figure 6. The
details of an individual algal production reactor are shown in
Figure 7. The procedure followed was to seed all five tubes

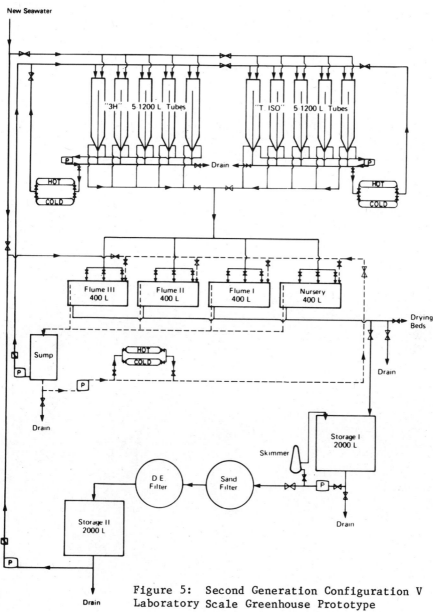

Figure 5: Second Generation Configuration V
Laboratory Scale Greenhouse Prototype

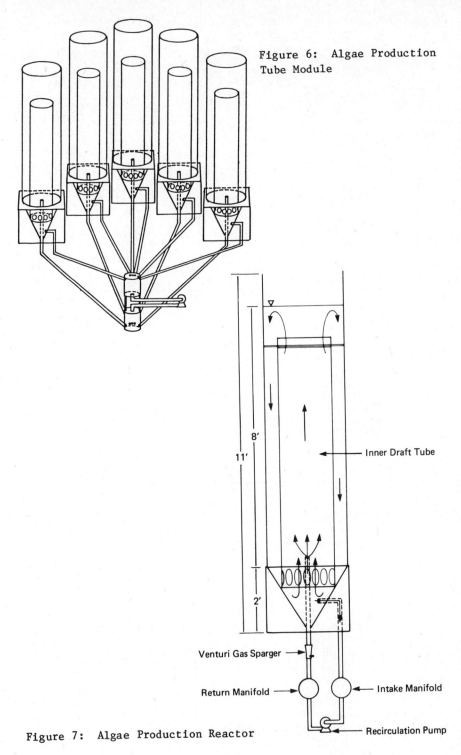

Figure 6: Algae Production Tube Module

8'

11'

2'

Inner Draft Tube

Venturi Gas Sparger →

Return Manifold →

Intake Manifold

Recirculation Pump

Figure 7: Algae Production Reactor

from laboratory grown cultures and to provide dissolved CO_2
and sunlight. Two or three doublings of cells occurred each
day. Algae were blended and fed to the oysters in the growing
tanks at night, holding one full tube to restock the other four
tubes each morning.

Oysters were grown in 400-liter baffled flumes as shown in
Figure 8. Water, after passing through the growing flumes, was
first passed through a protein skimmer, next a sand filter, and
finally a diatomaceous earth filter before being returned to
recharge the algae reactor modules. Heat exchangers permitted
the reactor temperature to be controlled. Automatic controls
maintained the desired pH and concentrations of CO_2 and O_2
(Thielker 1981).

This system was operated for over a year on a continuous
basis during 1980 and 1981. Data from this system is now being
analyzed, and a report will be completed in the near future.

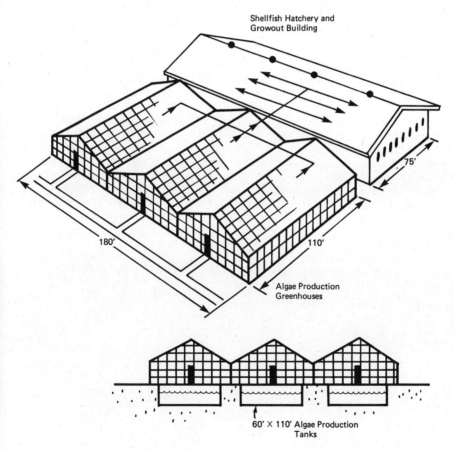

Shellfish Hatchery and
Growout Building

75'

180'   110'

Algae Production
Greenhouses

60' X 110' Algae Production
Tanks

Figure 8: Baffled Plume Design Flow Characteristics

Conclusions reached from constructing and operating these
laboratory systems are:
1. It is technically feasible to grow algae and bivalves in
   a recirculating system from nursery to adult size;
2. The nutritional requirements of bivalves are not yet well
   understood; and
3. A recirculating system is an important experimental tool
   to permit isolation of important problems and obtain
   useful solutions to them.

Economics

Thus far in this paper nothing has been said about economics
except that the eventual goal of the research is to produce
animals in captivity, with that degree of environmental control
necessary to insure uniform output of high quality disease-free
animals at a competitive price. At various points in the
research, commercial feasibility was estimated by applying the
"state-of-existing-understanding" to the design of a hypothetical commercial production system, or "bivalve factory." One
such design is shown in Figure 9. The cost of constructing and
operating these systems was estimated so that a price per bushel
of oysters produced could be estimated. In the 1976-1977 period
we estimated that oysters produced in such a facility could be
produced at prices competitive with those of "wild" oysters
harvested from boats (Figure 10) (Gaither 1976). Our optimism
proved to be premature, however, as the complexity and cost of
producing adequate food for the animals became more evident.

We are now confident that using a recirculating system
oysters can be spawned and grown as "singles" up to a size of
approximately 15mm diameter. Beyond that size the cost of algal
foods becomes uneconomical for continued growth in a recirculating system as configured in our laboratory. What we now have
is a system which offers several options as shown in Figure 11.

Conclusions

I want to conclude by observing that great progress has been
made through this research which can be applied to a variety of

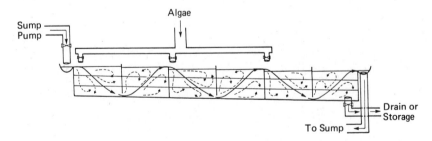

Figure 9: Commercial Concept I

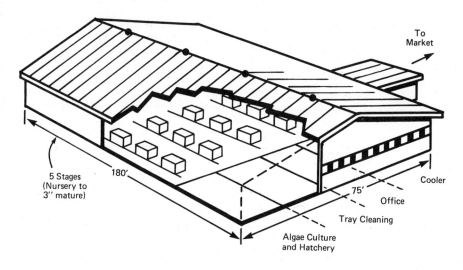

Figure 10:   Shellfish Building Commercial Concept I

problems. While we are not yet to the point that investors and
entrepreneurs are building and operating production systems with
the degree of process control we have achieved in our laboratory, what we have learned has been applied in several
commercial enterprises. Research yet needed to make cultivation
of marine bivalves in controlled environmental systems an
attractive commercial possibility includes:
1. Nutrition studies to understand the dietary requirements
   for both algae and artificial feeds to provide low cost
   food for bivalves
2. Selective breeding and hybridization of suitable bivalve
   species
3. Resolution of gas exchange problems in aquatic systems
   (Pruder and Bolton 1977a and b; Pruder 1981)
4. Genetic engineering applied to bivalve species improvement.

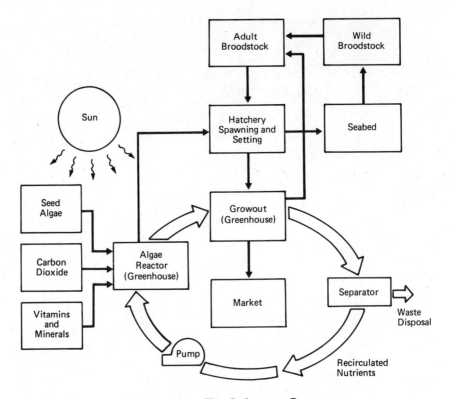

Figure 11: The Delaware Process

Attractive applications of information derived from this research include:

1. Bivalve hatcheries employing recirculating systems and grow-out in semi-controlled environments
2. Waste treatment systems employing aquaculture techniques which supply the biological oxygen demand and CO_2 for growing algae and bivalves, at least through the seed stage
3. Polyculture of marine animals and plants
4. Systems to produce biomass, particularly algae, for end uses of its own
5. Nutrient encapsulation techniques which can be employed for delivery of food and ingredients to a variety of commercially important marine species.

Like all research, the outcome cannot be predicted. We have, however, through this approach to an important problem, made significant progress, and the direction for further research seems clear.

Acknowledgements

I want to acknowledge with thanks the review and improvement of this paper by Dr. Ellis T. Bolton, Dr. Melbourne R. Carriker, Dr. Charles E. Epifanio, Dr. Christopher J. Langdon, Dr. Kent S. Price, Dr. Gary D. Pruder, Dr. Marenes R. Tripp, and Dr. Carolyn A. Thoroughgood.

References

1. Ali, S. M., 1981. Effects of inorganic particles on the growth of oysters, M.S. Thesis, University of Delaware, Newark, Delaware.

2. American Cyanamid Company, 1968. New engineering approaches for the production of Connecticut oysters, Stanford, Connecticut.

3. Carriker, M.R. and R.E. Palmer, 1979. Ultrastructural morphogenesis of prodissoconch and early dissoconch valves of the oyster Crassostrea virginica, Proceedings of Natural Shellfish. Assoc., 69:102-128.

4. Carriker, M.R., R.E. Palmer, and R.S. Prezant, 1980. Functional ultramorphology of the dissoconch valves of the oyster Crassostrea virginica, Proceedings of Natural Shellfish. Assoc., 70:139-183.

5. Epifanio, C.E., R.F. Srna, and G.D. Pruder, 1975. Mariculture of shellfish in controlled environments: A Prognosis, Aquaculture, 5:227-241.

6. Epifanio, C.E. and R.F. Srna, 1975. Toxicity of ammonia, nitrite ion, nitrate ion, and orthophosphate to Mercenaria mercenaria and Crassostrea virginica, Marine Biology, 33: 241-246.

7. Epifanio, C.E., C.M. Logan, and C.L. Turk, 1975. Culture of six species of bivalves in a recirculating seawater system, DEL-SG-1-76, College of Marine Studies, University of Delaware, Newark, Delaware.

8. Epifanio, C.E. and J. Ewart, 1977. Maximum ration of four algal diets for the oyster Crassostrea virginica (Gmelin), Aquaculture, 11:13-29.

9. Epifanio, C.E., 1979. Comparison of yeast and algal diets for bivalve molluscs, Aquaculture, 16:187-192.

10. Epifanio, C.E., in press. Phytoplankton and yeast as foods for juvenile bivalves; A review of research at the University of Delaware, Proceedings of Second International Conference on Aquaculture Nutrition; October 27-29, 1981, Lewes, Delaware.

11. Gaither, W.S., 1976. Testimony before Joint Subcommittee on Fisheries and Wildlife Conservation and the Environment and Subcommittee on Oceanography of the Committee on Merchant Marine and Fisheries, U.S. House of Representatives, 94th Congress, First and Second Sessions, p. 77-82; 95-102; 233-244.

12. Haskin, H.H., L.A. Stauber and J.P. Nackin, 1966. <u>Minchinia nelsoni</u> n. sp. (Haplosporida, Hapolosporidiidae): causative agent of the Delaware Bay oyster epizootic, Science, 153: 1414-1416.

13. Hicks, D., 1981. Sensory evaluation and selected chemical attributes of cultured and wild oysters, <u>Crassostrea virginica</u>, M.S. Thesis, University of Delaware, Newark, Delaware.

14. Langdon, C.J., in press. New techniques and their application to studies of bivalve nutrition, Proceedings of Second International Conference on Aquaculture Nutrition, October 27-29, 1981, Lewes, Delaware.

15. Muller, G.T., 1978. Effects of ration and temperature on growth rates and growth efficiency in the cultured oyster, <u>Crassostrea virginica</u>, M.S. Thesis, University of Delaware, Newark, Delaware.

16. Palmer, R.E. and M.R. Carriker, 1979. Effects of cultural conditions on morphology of the shell of the oyster <u>Crassostrea virginica</u>, Proceedings of Natural Shellfish. Assoc., 69:57-72.

17. Price, K.S. and D.L. Maurer, Eds., 1971. Artificial propagation of commercially valuable shellfish, Proceedings of a Sea Grant Conference, October 22-23, 1969, College of Marine Studies, University of Delaware, Newark, Delaware.

18. Pruder, G.D., 1976. Engineering aspects of bivalve molluscan mariculture: culture system configuration, DEL-SG-9-76, College of Marine Studies, University of Delaware, Newark, Delaware.

19. Pruder, G.D., E.T. Bolton, and S.F. Faunce, 1977. System
    configuration and performance bivalve molluscan mariculture,
    DEL-SG-1-77, College of Marine Studies, University of
    Delaware, Newark, Delaware.

20. Pruder, G.D., 1978. Effect of pH, carbon dioxide, and light
    on the growth of Thalassiosira pseudonana (Hustedt) Hasle
    and Heimodal clone 3H, an important food for bivalve
    molluscan mariculture, Ph.D. Dissertation, University of
    Delaware, Newark, Delaware.

21. Pruder, G.D., and E.T. Bolton, 1979. The role of CO_2
    enrichment of aerating gas in the growth of an estuarine
    diatom. Aquaculture, 17:1-15.

22. Pruder, G.D. and E.T. Bolton, 1979. Differences between
    cell division and carbon fixation rates associated with
    light intensity and oxygen concentration; implications in
    the cultivation of an estuarine diatom, Marine Biology,
    59:1-6.

23. Pruder, G.D., 1981. Aquatic production systems: algae, In
    Proceedings, International Workshop on Nursery Culturing of
    Bivalve Molluscs, State of University Ghent, Belgium, p.
    219-226.

24. Pruder, G.D. and J. Ewart, in press. Role of organically
    coated silt on the growth of oysters.

25. Romberger, H.P. and C.E. Epifanio, 1981. Comparative
    effects of diets consisting of one or two algal species upon
    assimilation efficiencies and growth of juvenile oysters,
    Crassostrea virginica (Gmelin), Aquaculture, 25:89-94.

26. Sick, L.V. and C.C. Johnson, 1977. Nutrient and trace metal
    concentrations in the abiotic and biotic components of a
    recirculating seawater culture system for oysters, American
    Zoologist, 17(4):977-78.

27. Sick, L.V., 1981. Effects of the ambient environment in
    metabolic regulation of shell biosynthesis in marine bivalve
    molluscs, In Proceedings of the Second International
    Conference on Aquaculture Nutrition, Rehoboth Beach,
    Delaware.

28. Thielker, J.L., 1981. Design and test operation of an
    intensive controlled-environment oyster production system,

DEL-SG-07-81, College of Marine Studies, University of Delaware, Newark, Delaware.

29. Thoroughgood, C.A., 1979. Evaluation of palatability and selected nutrients of bivalve molluscs grown in controlled mariculture and natural systems, In Proceedings World Mariculture Soc., Honolulu, Hawaii.

Application of Recombinant DNA Technology to the Domestication of Marine Plants

John N. Vournakis

Syracuse University
Syracuse, New York

Abstract

A major underexploited worldwide resource for crop growth
and aquaculture is the coastal zone region. Two factors that
contribute to this situation include (1) the limited availability of food-producing plants having osmoregulatory mechanisms
that allow them to survive the stress imposed by the salinity of
the marine environment; and (2) the increasing pollution of
coastal waters resulting in the increase in concentration of
toxic substances, including deleterious heavy metals. Recent
advances in molecular biology and genetic engineering technology
have made it possible to isolate, manipulate, and study the
expression of specific eukaryotic genes. In principle, it is
now possible to engineer new species of plants with permanently
altered genetic properties. This paper presents a review of the
current status of recombinant DNA technology and proposes two
specific applications of these techniques to create new marine
plants having desirable characteristics. The first set of
studies suggested would expand the current understanding of
osmoregulation. A fruit-producing salt tolerant hybrid tomato
has been previously constructed by classical genetic crossing of
the commercial non-salt-tolerant species Lycopersicon with the
inedible but salt-tolerant species Lycopersicon. Experiments
are proposed that involve the preparation of cDNAs from the
purified mRNAs of the hybrid and wild-type plants and their use
to probe the cloned genomic DNA libraries of these plants in
differential screening studies that can identify the genetic
elements involved in conferring the osmoregulatory trait. A

gene cloning system in plants using the <u>Agrobacterium</u> Ti (tumor inducing) plasmids as vectors is described, and a strategy is discussed for producing new species having enhanced salt tolerance. The second set of studies focuses on the issue of producing strains of algae and phytoplankton, important marine animal food sources, that have increased tolerance to heavy metal contamination. Recent work on the regulation of the gene for the cadmium binding metallothionein protein in mammalian systems is presented. The use of the SV40 vector system to allow the cloning of specific genes and the study of their regulation is discussed. The Cd^{++} binding protein gene is viewed as a model system in a discussion of the experimental information and requirements needed for generating new marine plants that can better tolerate heavy metal toxicity, and for generating novel biological pollution control devices.

Problems of Crop Cultivation in the Coastal Zone

The coastal zone regions of the world encompass a major underutilized water resource of enormous potential for crop growth and aquaculture (Epstein and Norlyn 1977). During the past decades these regions have been abused by a massive onslaught of pollutants. This pollution is often so severe that plant and animal life becomes impossible due to the accumulation of heavy metal and organic chemical wastes. Successful utilization of the coastal zone requires serious efforts to design new strategies for removing deleterious industrial contaminants. In parallel with these efforts, progress in the development of food-bearing plants that can flourish in the highly saline marine environment must occur. Only a small number of plants currently exist that are both salt tolerant and produce edible fruit. Most plants of agricultural value do not have the osmoregulatory mechanisms needed to allow their cells to survive the osmotic pressure gradients imposed by the concentration of salt in the sea. In addition, inland regions that are heavily irrigated, such as the highly productive central valleys of California, are becoming increasingly saline (Norlyn 1979). There is, therefore, a serious need for a research program aimed at creating plants capable of survival in coastal waters, as well as in saline soils on land. The molecular mechanism of osmoregulation in plants is not understood (Jeffries 1979). Although there has been progress in identifying the chemical components in cells of marine plants responsible for the plants' salt-tolerance (Jeffries 1979), little is known about the regulation at the genetic level of the metabolic pathways that produce either the organic osmotica or the specialized ion transport systems that confer salt tolerance. It is not known if all plants have the genetic potential to osmoregulate, nor how the osmoregulatory genes are induced, linked, regulated, etc. Answers to these and many similar questions will help pave the way for the development of salt-tolerant plants of many varieties.

Recent advances in molecular biology and genetic engineering technology have made it possible to isolate, manipulate, and study factors relevant to the regulation of the expression (transcription and translation) of eukaryotic genes (Abelson 1980). It has become possible to introduce foreign genes into plants so as to engineer new species with permanently altered genetic traits (Otten et al. 1981). This paper presents a review of some aspects of modern genetic engineering technology and proposes two specific applications of these methods to create marine plants having desirable characteristics.

The first set of studies describes an approach whereby recombinant DNA can be used to study osmoregulation in an edible fruit-producing tomato plant that has been bred by classical genetic methods to survive in seawater (Norlyn 1979 and Rick 1972).

The second example focuses on the development of species of marine plants that have an enhanced tolerance to heavy metal contamination. This would require an understanding of the regulation of genes coding for proteins that bind heavy metals, such as the metallothioneins (Kagi and Nordberg 1979; Webb 1979), and the development of appropriate cloning systems so that heavy metal tolerance could be engineered into marine plants. These studies could provide a basis for the development of new pollution control devices.

Genetic Engineering and Recombinant DNA
Genetic engineering has been defined by Professor David Baltimore of MIT as "the insertion of new genes or the replacement of defective genes in cells of higher organisms." Genetic engineering can be achieved two ways: either so that the genes of cells in a single individual or one individual in a population are changed, or so that the altered but improved genome is transferred to all the individuals in the population and thereby to all future generations. In the first type of genetic engineering the offspring of the individual do not have an altered genetic constitution. This should be referred to as gene therapy for the individual. Whereas in the second type, the defective gene is eliminated from the population.

The recombinant DNA (rDNA) method is one of many biochemical technologies used to perform genetic engineering operations. It involves the in vitro recombining of DNA molecules and the introduction of these recombinant molecules into living cells where they replicate and become expressed as gene products. The rDNA method, according to Morrow (1979), accomplishes the following: it allows the isolation of pure DNA sequences in biochemically meaningful amounts; it facilitates the determination of the nucleotide sequences of genes from which protein sequences can be derived using the genetic code; it makes studies of gene regulation possible at the nucleotide sequence level, e.g. via site-directed mutagenesis, nucleotides in and near genes that affect the efficiency of gene expression

can be identified; and it provides a biochemical approach
whereby the biosynthesis of large amounts of specific gene
products can be accomplished for study and, potentially, for
practical applications in medicine, agriculture, and many other
areas important to society.

The DNA method can be used only when a combination of
several basic systems is available for the biological system of
interest. These requirements (Table 1) include the following:
(1) a system for the transport of a foreign gene into a host
organism, i.e. a DNA vehicle or vector capable of being
replicated in the host cell type. This vector, typically a
plasmid, virus or bacteriophage, will carry the foreign DNA into
the cells of a living organism and will have an origin of
replication so that it will be synthesized along with the
foreign DNA and amplified in the host cell; (2) availability of
the foreign DNA in biochemically meaningful quantities; (3) a
method for the specific joining of the foreign and vector DNAs
to generate the rDNA molecule of interest; (4) a transforming
system for each host species, i.e. a specific experimental
protocol for introducing the rDNA into the approriate host
organism so that it carries and expresses the genotype present
on the rDNA and must be developed for each host species; and (5)
a method for the rapid screening of the potentially transformed
host cells to identify the ones carrying the rDNA genotype (i.e.
the ability to select for the correct clones). If the above
requirements can be satisfied in a single system, the recombinant DNA method can be applied to achieve one or more of
Morrow's goals stated above.

Three Landmark Biochemical Breakthroughs

The development and serious application of the recombinant
DNA method to problems of gene structure and regulation in
prokaryotic and eukaryotic organisms were pre-dated by three
remarkable biochemical advances within the past twenty years.

Table I. Components Neccesary for Recombinant DNA Method

---

1. A DNA vector (vehicle) capable of being replicated in the
   host organism.

2. A foreign DNA molecule (e.g. a gene).

3. A method for joining the foreign DNA with the vector to
   generate the desired recombinant DNA molecule (rDNA)

4. A method for transforming the host organism with the rDNA

5. A method for screening the transformants to isolate the
   correct clone containing the foreign DNA.

These breakthroughs include: the discovery of the restriction enzymes, endonucleases isolated from prokaryotes that can cleave double-strand DNA in a highly specific manner at precise sequences; the discovery of reverse transcriptase, an enzyme that can be used to catalyze the biosynthesis of DNA from RNA templates so that complementary DNA (cDNA) copies of purified messenger RNA (mRNA) molecules can be prepared; and the development of high resolution, rapid methods for obtaining DNA and RNA nucleotide sequences. In addition, the discovery, purification, and characterization of a number of other nucleic acid enzymes such as RNA and DNA ligases (Higgins and Cozzarelli 1979; Gumport and Uhlenbeck 1981), RNA and DNA endonucleases (Simpson 1981; Vournakis et al. 1981), RNA and DNA polymerases (Kornberg and Gefter 1971; Roeder 1976), etc. have occurred during the same period and have contributed significantly to the set of biochemical tools currently in use for rDNA research.

There is extensive literature describing the discovery, physiological significance, purification, and characterization of restriction enzyme (Colowick and Kaplan 1980). Figure 1 illustrates the sequence specificity of the restriction enzyme EcoRl. Two different DNA sequences, each having an EcoRl site, can be recombined, following cleavage by EcoRl, with DNA ligase to make an rDNA molecule. Figure 1 represents the general principle that restriction enzymes are a powerful and necessary tool in the process by which recombinant DNA molecules are

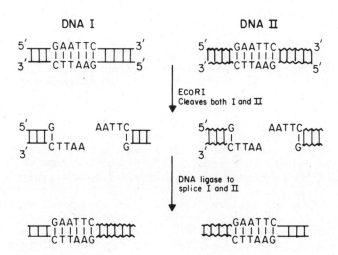

Figure 1. Cleavage of DNA by the restriction enzyme EcoRl and subsequent religation with DNA Ligase to form recombinant DNA molecules.

constructed. More than 200 restriction enzymes are currently
available (Colowick and Kaplan 1980).

The importance of the discovery of reverse transcriptase
(Baltimore 1970) to rDNA technology is illustrated in Figure 2.
This enzyme is used to catalyze the copying of a highly purified
eukaryotic mRNA into cDNA as the first step in the in vitro
synthesis of a "cDNA gene." The reverse transcriptase will
catalyze the polymerization of a DNA molecule from an RNA
template, provided that there is a small DNA primer hybridized
to the RNA. In this case (Figure 2) the primer used is
synthetic oligo $(dT)_{15}$, a homopolymer of thymidylic acid,
which hybridizes to the 3'-polyadenylate sequence of the mRNA.
The enzyme then directs the extension of the DNA primer in a 3'
to 5' direction along the mRNA to make a single-stranded DNA
complementary copy of the messenger molecule, provided that the
correct concentrations of deoxynucleotide triphosphates are
present in the reaction mixture. The complete snythesis of the
double-stranded cDNA gene requires several additional enzymatic
steps, as illustrated. cDNA genes can be recombined with
appropriate vectors and cloned in appropriate hosts to obtain
sufficient amounts for sequencing studies (Efstratiadis 1976),
or for use as probes in experiments designed to isolate the
specific natural genes from the genomic DNA of an organism
(Lautenberger 1981). cDNA genes have also been recombined with
vectors specifically designed to have high efficiency promoters
that result in the expression of those genes to produce proteins
in large quantity in the particular host organism. The
development of strains of E. coli that overproduce α-interferon
proteins from cloned α-interferon cDNA genes is one example of
the utilization of this approach to produce large amounts of an
economically important substance (Goeddel et al. 1980).

The third major biochemical advance, the development of
techniques for the rapid sequence analysis of DNA and RNA, was
developed simultaneously by Maxam and Gilbert (1977) at Harvard,
and by Sanger et al. (1977) at Cambridge. The ability to clone
specific large segments of DNA is an impressive advance.
However, without the advent of these new sequencing methods, it
would be impossible to extract vital information that exists in
the DNA and that enhances our understanding of the gene structure and regulation of gene expression. The DNA sequencing
method of Maxam and Gilbert is essentially a chemical method
that relies on a strategy involving labeling one end of a DNA
molecule with the radiosotope ^{32}P and performing in parallel a
set of partial chemical reactions, each specific for one of the
four nucleotide bases, resulting in the removal of that base
from the chain. The nested set, i.e. the full-length and all
possible partial length end-labelled oligomers, thus generated
are separated by high resolution polyacrylamide gel electrophoresis, and the sequence can be read following autoradiography
to locate the position of the bands on the gel. An example of a

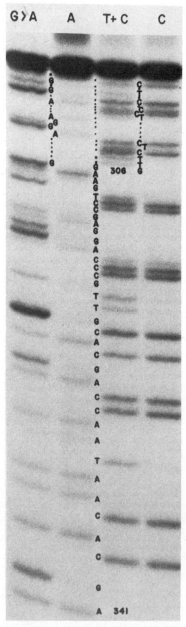

Figure 2. Reverse transciptase directed synthesis of a cDNA gene. Steps following ss-cDNA synthesis require the enzymes DNA polymerase I and S1 nuclease.

sequencing gel is shown in Figure 3. This is a portion of the
rabbit β-globin gene sequence obtained by Efstratiadis et al.
(1976).   RNA can be sequenced using a similar strategy except
that either base-specific ribonucleases or chemicals can be used
to generate the information (Simpson 1981; Vournakis et al.
1981).   These methods can have the power to allow a single

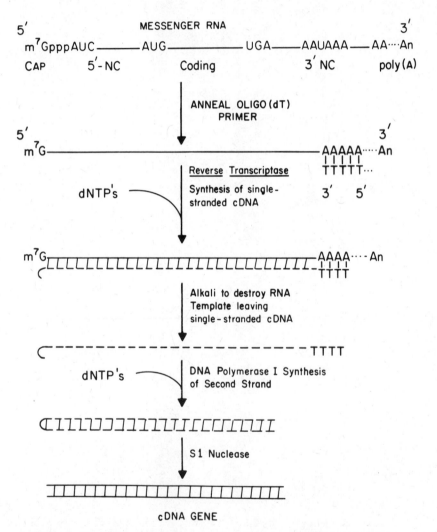

Figure 3. DNA sequencing gel showing a portion of the
sequence of the rabbit β-globin cDNA gene, from Efstratiadis et
al. (1976).

worker to sequence up to 10,000 basepairs (bp) of DNA per year, whereas it was possible to obtain the sequence of only very short DNA molecules (10 to 20 bp) by classical methods.

<u>Cloning Vectors</u>

An increasing number of cloning vectors capable of transforming a variety of organisms is becoming available. Plasmids and bacteriophages are the most commonly used vectors in prokaryotic cloning systems. Certain of the plasmids and phages have been engineered so as to include convenient restriction enzyme sites into which foreign DNA can be cloned, the plasmid origin of replication, and genes coding for selectable traits. A commonly used vector, plasmid pBR322 (Suttcliffe 1978), is illustrated in Figure 4. It is a covalently closed circular DNA molecule consisting of 4,362 basepairs and has been completely sequenced (Suttcliffe 1978). It also includes two drug resistance markers, the ampicillin and the tetracycline resistance genes, an origin of replication, and several restriction enzyme sites at useful loci within the molecule. Transformation of the appropriate host <u>E. coli</u> strain with pBR322 generates resistance to both ampicillin and tetracycline. These cells can be selected by growing them in media that contain the two drugs. If one, however, clones a segment of foreign DNA into the single restriction enzyme site for Pst I the ampicillin resistance is lost since the gene is interrupted. Thus, a simple strategy emerges for using a well-engineered plasmid vector to clone in <u>E. coli</u>.

Recently, considerable progress has been made in discovering and developing plasmid vectors for the Gram-positive bacteria

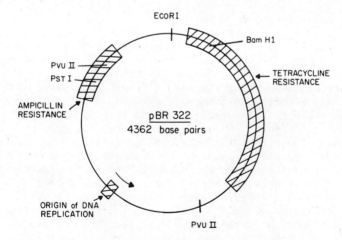

Figure 4. The commonly used plasmid vector pBR322. Note that it contains two drug resistance markers, and several restriction enzyme sites are indicated.

such as <u>Bacillus</u> <u>subtilis</u> (Lovett and Keegins 1979). Combination or <u>shuttle</u> vectors have also been constructed that can
transform two organisms, e.g. both <u>E. coli</u> and <u>B. subtilis</u>
(Messing, Crea, and Seeburg 1981). These shuttle vectors are
derived from segments of plasmids that transform each organism
and have the replication origin and one or more drug resistance
markers from each of the original plasmids. Thus, they can
replicate in and thereby transform both organisms.

     Bacteriophages are also commonly used vectors. The λ phages
are capable of carrying rather large segments of foreign DNA
(approximately 10,000 base pairs) compared to plasmids, and are
therefore particularly useful for constructing libraries of
genomic DNA (Williams and Blattner 1980). The M-13 single-
stranded phage system is very useful for cloning DNA segments
for sequence analysis using the Sanger method mentioned above
(Messing, Crea, and Seeburg 1981).

     Eukaryotic cloning vectors include small viruses such as
SV40 (Hamer 1980) which are particularly useful in experiments
designed to study gene regulation in mammalian cells (Pavlakis
et al. 1981). The Ti (tumor inducing) plasmids of <u>Agrobacterium</u>
<u>tumefaciens</u> are vectors that can transform plant cells and are
discussed in detail below. Eukaryotic/prokaryotic shuttle
vectors are currently available. The prokaryotic part of such a
shuttle vector (typically part of pBR322) allows for the large
scale preparation of the vector (and any genes cloned into it)
in <u>E. coli</u>. The eukaryotic aspect of the vector (e.g. the
relevant parts of the SV 40 genome) confers the functions that
are vital to being able to transform eukaryotic cells. It can
be used, therefore, to study gene regulation in mammalian cells
(Pavlakis et al. 1981).

Combining Components to Clone Genes

     The various biochemical components required for recombinant
DNA studies listed in Table I and discussed above are combined
to perform gene cloning experiments. Figure 5 illustrates the
steps involved in cloning in the prokaryote, <u>E. coli</u>. The need
for restriction enzymes, ligases, a transformation protocol, a
vector, and a foreign DNA segment are shown. Figure 6 describes
the techniques for the preparation and cloning of a eukaryotic
cDNA gene in <u>E. coli</u>. Such a protocol would be used to prepare
large amounts of the eukaryotic cDNA genes for either sequence
analysis or for use as probe in experiments designed to isolate
the natural gene from the chromosomal DNA of the organism.
Figure 6 illustrates the key role of the reverse transcriptase
enzyme which is an absolute requirement for such studies.

     The remainder of this paper focuses on the application of
the recombinant DNA methodology described above to two efforts:
attempting to understand osmoregulation in plants at the
molecular level in order to use such an understanding to develop
new salt-tolerant species; and studying the mechanism of
expression of genes coding for heavy metal binding proteins as

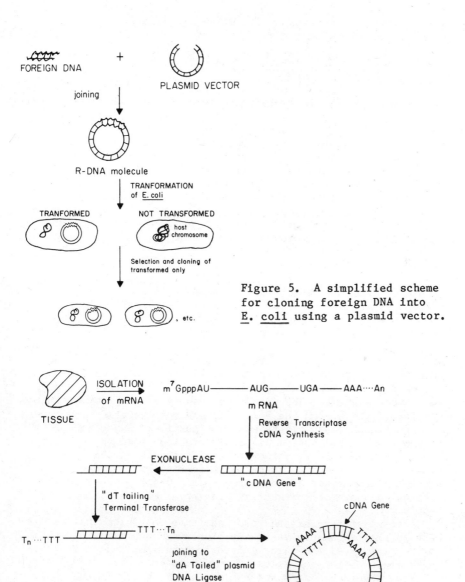

FOREIGN DNA   +   PLASMID VECTOR

joining

R-DNA molecule

TRANFORMATION
of E. coli

TRANFORMED   NOT TRANSFORMED

host
chromosome

Selection and cloning of
transformed only

, etc.

Figure 5. A simplified scheme
for cloning foreign DNA into
E. coli using a plasmid vector.

TISSUE

ISOLATION
of mRNA

m⁷GpppAU——— AUG——— UGA——— AAA····An

mRNA

Reverse Transcriptase
cDNA Synthesis

EXONUCLEASE

"cDNA Gene"

"dT tailing"
Terminal Transferase

Tₙ···TTT

TTT···Tₙ

joining to
"dA Tailed" plasmid
DNA Ligase

cDNA Gene

AAAA
TTTT
TTTT
AAAA

Transformation, selection and
cloning in host

Figure 6. The cloning of a cDNA gene via plasmid
transformation into a prokaryotic host organism.

91

the necessary first step in designing saltwater plants that
tolerate and accumulate high levels of heavy metals.

Development of a Salt-Tolerant Tomato Plant by Genetic Engineering

As an introduction to the problems involved in attempting to
create new salt-tolerant plant species by genetic engineering it
is important to focus on certain physiological properties of
plants. Algae and most active organs of plants, including
leaves and roots, are 85-95 percent water by weight (Norlyn
1979). Plants can be safely desiccated or frozen for short
periods of time. However, water is vital in the normal func-
tioning of plants. Water is abundant on earth. Seventy-one
percent of the surface of the globe is kept supplied by water
via the hydrological cycle. But despite its relative abundance
compared to other nutrients, water is acquired by plants at a
great metabolic cost through a complicated structural and
functional adaptation (Epstein and Norlyn 1977; Norlyn 1979).
There are two principal reasons why this is, in energetic terms,
expensive. First, in order to compensate for the concentration
of sodium chloride--which is approximately 0.5 molar--and high
concentrations of other salts in the oceans, marine algae and
other marine plants must maintain an intracellular concentration
of some solute, not necessarily salts, capable of preventing
osmotic desiccation. Second, plant species that grow and thrive
on land must cope with the solid matrix of the soil in order to
acquire water and nutrients through a root system, while
presenting leaves to the atmosphere for obtaining light energy
and carbon dioxide. But water can evaporate from leaves, thus
drawing moisture from the soil. If the soil is arid or water is
in short supply, the plants will not survive. So, plants in the
oceans and on land have a difficult problem obtaining water for
survival. There is recent evidence (Norlyn 1979) that the
percentage of saline water on Earth is rising. The data
strongly infers that salinity looms as a major problem in the
chemical economy of terrestrial plants as well as marine
plants.

A consideration of the major adaptive mechanism that allows
certain plants to survive in the highly saline marine
environment, i.e. the osmoregulatory mechanism, indicates that
there are two distinct modes of physiological adaptation:

Some plants can grow and thrive in salt water because their
cells are able to transport salts and come to osmotic
equilibrium with their environment by having a high internal
salt concentration. This form of adaptation implies that the
plant's biochemical machinery, i.e. its metabolic pathways,
protein synthetic system, etc., can function adequately at salt
concentrations much higher than normal for terrestrial organisms.

Other plants osmoregulate by producing and maintaining high
intracellular concentrations of certain organic solutes, known
as organic osmotica (Jeffries 1979). These plants apparently

have evolved metabolic pathways that generate innocuous organic
solutes (e.g. proline, sorbitol, etc.) that accumulate in the
plant cell. These solutes are localized, usually, in the plant
cell cytoplasm, and allow the internal milieu of the plant cell
to be at osmotic equilibrium with the cell wall and vacuole
(Jeffries, 1979), which generally maintain high salt concentrations in marine plants. Thus, osmotic disruption of the plant
is avoided. The detailed biochemical mechanisms that confer the
osmoregulatory property by either of the above mechanisms are
not understood at the molecular level. It is likely that the
class of plants that accumulate salts have evolved ion transport
systems having properties quite different from the ion transport
proteins in terrestrial plants. The plants that accumulate
organic osmotica have evolved metabolic pathways that synthesize
the organic solutes. Interesting questions regarding the
control of these pathways by the external ionic environment
invite study.

Salt Tolerant Tomato Plants from Classical Genetics
The work of Kelly et al. (1979) and Norlyn (1979) has
resulted in the development of a tomato plant that produces
edible fruit and survives in seawater. This system is used as a
paradigm to illustrate the potential of genetic engineering in
the study and development of salt-tolerant plants. Figure 7

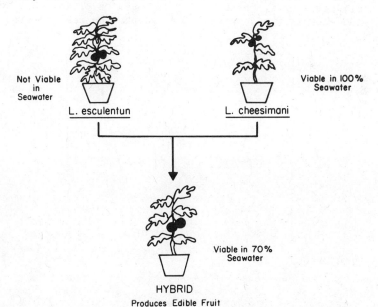

Figure 7. Genetic hybridization of the tomato species
Lycopersicon esculentum and Lycopersicon cheesimani produces a
hybrid with viability in 70 percent seawater that produces
edible fruit.

illustrates the use of classical genetic techniques (cross
pollination) and selection on the basis of two criteria,
survivability and fruit production, to produce an interesting
hybrid. Lycopersicon esculentum, a terrestrial and agricul-
turally important tomato species, was crossed with Lycopersicon
chessimani, a salt-tolerant tomato which produces dwarfed and
inedible fruit, to yield a viable productive hybrid that can
survive in up to 70 percent seawater. Apparently, the hybrid
inherited the osmoregulatory traits necessary for the viability
of the stem and root system in high salt from L. chessimani, as
well as the traits for generating fruit with cells that do not
accumulate salt and with desirable growth characteristics from
L. esculentum.

Some data relevant to the properties of the original strains
is shown in Figure 8 taken from Epstein and Norlyn (1977). L.
esculentum, which doesn't survive at high salt concentration is
seen to be disrupted by 0.5x seawater. The sodium breaks into
the root cells at the high salt concentration and the plant
dies. L. chessimani continues to accumulate sodium beyond 0.7x
seawater and can accumulate as much as 6 percent of its dry
weight as salt in its leaves.

The hybrid illustrated in Figure 7 grows well in up to 70
percent seawater and at all lower concentrations. This
indicates that it has the ability to osmoregulate, but not to
the same degree as L. chessimani. It would be of interest to
determine whether the osmoregulatory adaptation results from a

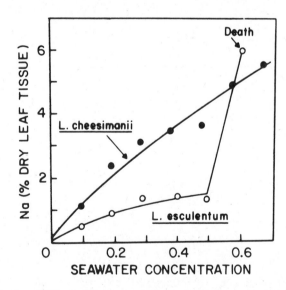

Figure 8. Effect of exposure to varying concentrations of
seawater on the accumulation of sodium and on the viability of
two tomato species.

permanent alteration in the hybrid or if it is an inducible
phenomenon responding to the amount of salt in the external
environment in a quantitative fashion. It would, therefore, be
necessary to attempt to isolate the genes involved in conferring
the osmoregulatory properties. Studies of these genes might
provide a basis for understanding the osmoregulatory phenomenon
and would provide the basis for developing a strategy whereby
such genes might be cloned into other species to create new
strains of food-bearing plants having salt tolerance.

Cloning and Expression of Foreign Genes in Plants

In order to proceed with a potentially useful study of
osmoregulation in tomato plants, it is important that a system
for cloning foreign genes into plants be described. The best
developed model to date is the Agrobacterium tumefaciens Ti
plasmid system that has been used to clone foreign genes into
tobacco plants (Schell and VanMontagu 1977). Agrobacterium
tumefaciens is a Gram-negative soil bacterium that infects
dicotyledonous plants and produces a neoplastic growth called a
crown-gall tumor. The cells from such a plant tumor can in some
cases be used to regenerate whole plants (Braun and Wood 1976).
These crown gall tumors result from the transfer of DNA from the
microbe to the plant. The plasmids that perform this unusual
function are known as Ti (tumor-inducing) plasmids. They carry
genes for the biosynthetic pathway of the opines, notably
octopine and nopaline, which are basic amino acid derivatives of
arginine and lysine. The bacterium is unable to synthesize the
opines, but prefers those amino acids as a primary carbon
source. Figure 9 illustrates the unique cycle whereby the A.
tumifaciens infects the plant with the plasmid and grows in the
soil using the plant-synthesized opines.

The Ti plasmids are very large closed circular DNA molecules
150-200 kilobases (kb) in size. They contain genes which encode
many properties such as the oncogenic trait (i.e. the induction
of the crown gall tumor), opine biosynthesis, antibiotic
resistance, etc. Figure 10 illustrates a typical Ti plasmid.

Plant cells transformed with the Ti plasmid have a small
segment of the plasmid DNA, the 20 kb tDNA, stably integrated
into their genome (Otten et al. 1981). The tDNA segment can,
therefore, be used as a cloning element to carry genes of
interest into the plant genome where it remains and can be
expressed in all future generations. This strategy is shown in
Figure 11. It is known that tDNA is often present as tandem
repeats in as many as five copies per cell (Chilton 1979). One
therefore constructs a new Ti plasmid with a gene of interest
cloned into the tDNA portion and transforms to obtain a new
healthy plant with the new trait. The tDNA insertion into the
plant DNA is accomplished either by being cleaved into a linear
molecule prior to integration, or by integrating as a circle
(Zambryski 1980). There has been considerable success in
cloning and expression of foreign genes into plants using the Ti

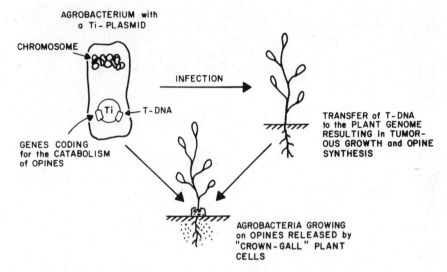

Figure 9. Symbiotic relationship between <u>Agrobacterium</u> <u>tumefaciens</u> and dicotyledenous plants illustrating the role of the Ti plasmids and their tDNA element.

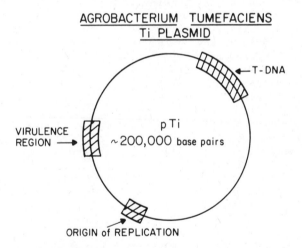

Figure 10. Diagram representing some common aspects of Ti plasmids.

plasmid system. For example, the rabbit β-globin gene has been expressed in tobacco plant cells.

In summary, there exists a well described system available for the genetic engineering of plants, and for using recombinant DNA methods to study gene expression in plants. This system can be used in the experiments proposed below for studies of osmo-

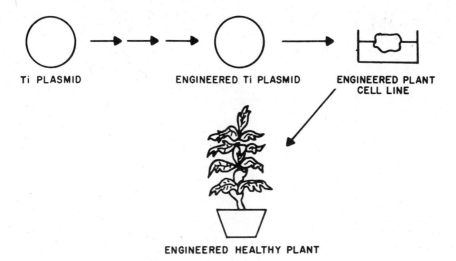

Ti PLASMID   ENGINEERED Ti PLASMID   ENGINEERED PLANT
                                                           CELL LINE

ENGINEERED HEALTHY PLANT

Figure 11. A scheme for genetic engineering in plants, using Ti plasmids as vectors.

regulation, toward developing a strategy for creating new plant species that will thrive in seawater.

A Study of the Molecular Biology of Osmoregulation

The hybrid tomato that grows in 70 percent seawater, discussed above, can be used in a series of experiments to study the molecular mechanism of osmoregulation using rDNA techniques. The genes coding for the osmoregulatory trait are clearly present in the hybrid. However, they may be inducible genes, i.e. they are, perhaps, not expressed when the hybrid is grown in fresh water and induced when it is grown in 70 percent seawater. If so, it should be possible to identify the genes involved by using the strategy illustrated in Figure 12. Messenger RNA populations from induced and uninduced plants are isolated and differential comparative gel electrophoresis is used to partially purify the induced mRNAs. cDNA copies of these mRNAs are prepared and are used as probes to purify the osmoregulation genes from the chromosomal DNA of the plants by screening gene libraries prepared in advance. A somewhat more complicated version of this study is shown in Figure 13. This alternative approach will work if the genes in question are not inducible by the presence of high salt in the environment. The experiments are relatively straightforward and involve much of the rDNA technology discussed earlier. Studies of this type would have a reasonable probability of advancing the knowledge of the genetic traits in question, and would provide the fundamental basis for future introduction of the salt tolerance trait into other plants. This would be accomplished by use of plant

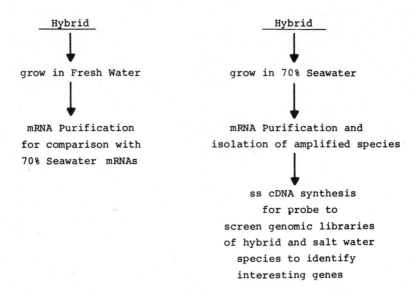

Figure 12. A strategy for isolation of genes involved in osmoregulation in plants using the hybrid tomato plant depicted in Figure 7.

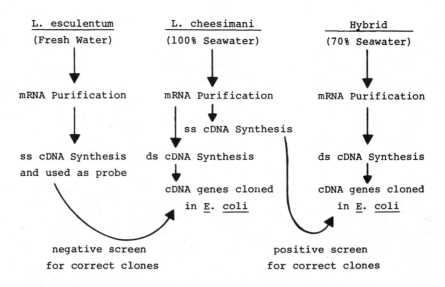

Figure 13. A complicated strategy for cloning osmoregulatory genes using the Lycopersicon esculentum, Lycopersicon chessimani, hybrid system described in the text.

cloning vectors, such as the Ti system described above.

Epstein and Norlyn (1977) point out that in those cases where plants tolerate salt water by accumulating high internal ion concentrations, e.g. barley, which accumulates potassium ions, there is evidence that the membrane ion transport protein levels and types change in response to the external ionic environment. It is possible that in such cases a relatively small number of genes would be responsible for the osmoregulation. If so, the genetic engineering approach would have a better chance of being effective.

Development of Heavy Metal Binding Protein Genes as Potential Pollution Control Devices

The metallothioneins are small cysteine-rich proteins that have a very high affinity for heavy metal ions such as zinc, cadmium, copper and mercury (Kagi and Nordberg 1979; Webb 1979). Recent interest in these proteins has focused on their role in zinc and cadmium metabolism and in heavy metal detoxification, primarily in mammalian organisms (Webb and Cain 1982). The biosynthesis of these proteins in mammalian cells can be induced by injecting high levels of the heavy metal ion into the bloodstream of the animal (Webb and Cain 1982). The metallothioneins represent a scavenger system, binding the metal ion in the bloodstream and providing an exit mechanism for removal of the complex via the kidneys.

Genes coding for the metallothioneins are present in all mammals, and genes coding for other similar heavy metal binding proteins are found throughout the phylogenetic hierarchy, even in prokaryotes. Some of these genes are receiving considerable attention in studies using rDNA techniques. Palmiter (Glanville, Durnam, and Palmiter 1981) has cloned the cadmium binding metallothionein I gene and has obtained its complete nucleotide sequence. He has discovered that this gene has putative control sequences upstream from the coding region that are similar to those found on other mammalian genes. Figure 14 illustrates this gene and indicates the existence of a region containing a sequence TATAAA, which is required for the accurate transcription of the gene. Studies by Hamer and Walling (1982) and Palmiter (Mayo, Warren, and Palmiter 1982), using SV40 vectors and cloning the metallothionein I gene into monkey kidney cells, are aimed at delineating the absolute sequence requirements for induction and regulation of the gene. These authors demonstrate that the expression of the entire Palmiter gene sequence can be regulated in the transformed mammalian cell line by including Cd^{++} in the medium. The effects on this regulation of systematically removing regions upstream from the coding sequence in order to identify the exact limits of the gene are being explored.

Such studies and many others focusing on metal-binding protein genes from other species can form the basis for the potential development of industrial products or, in the future,

METALLOTHIONEIN - I GENE

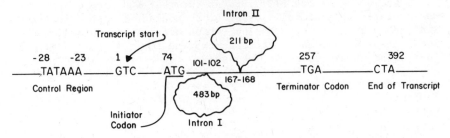

Figure 14.   A sketch of the metallothionein-I, cadmium
binding protein, gene from the work of Palmiter et al. (1982).
The putative transcriptional control region ("TATA" box) is
shown at positions -23 to -28 from the origin of transcription.
The initiator and terminator codons are indicated as are the two
introns.

salt tolerant plants as pollution control devices.   If large
quantities of pure metallothionein proteins can be made by rDNA
methods, then it should become possible to produce affinity
column resins containing these proteins bound to solid
supports.   Such resins might be used in sewage treatment plants
or in industrial processes, e.g. wineries and breweries, to
remove cadmium, mercury, etc. from the water.   The type of
genetic engineering of plants described above could also be used
to clone metallothionein genes into species of marine plants
that could then be strategically grown in harbors and coastal
waters to provide a natural system for reducing toxic metal
concentrations.   The achievement of these goals is far off, but
the path is now open.

References

1. Abelson, J., 1980. Science, 209:1319.

2. Baltimore, D., 1970. Nature, 226:1209.

3. Braun, A.C. and H.N. Wood, 1976. Proc. Nat. Acad. Sci. USA, 73:496.

4. Chilton, M., 1979. In D.W. Rains, R.C. Valentine and A. Hollaender, Eds., Basic Life Sciences, Genetic Engineering of Osmoregulation, 14:23.

5. Colowick, S.P. and N.O. Kaplan, Eds., 1980. Methods in Enzymology, Nucleic Acids, Vol. 65, Academic Press, New York, NY.

6. Efstratiadis, A., F.C. Kafatos, A.M. Maxam and T. Maniatis, 1976. Cell, 1:279.

7. Epstein, E. and J.D. Norlyn, 1977. Science, 197:249

8. Glanville, N., D. M. Durnam and R. D. Palmiter, 1981. Nature, 292:267.

9. Goeddel, D.V. et. al., 1980. Nature, 287:411.

10. Gumport, R.I. and O.C. Uhlenbeck, 1981. In T. Papas and J. Chirikjian, Eds., Gene Amplification and Analysis, Vol. 2, Structural Analysis of Nucleic Acids, Elsevier–North Holland, New York, p. 314.

11. Hamer, D.H., 1980. In J.K. Setlow and A. Hollaender, Eds., Genetic Engineering, Plenum, New York, p. 83.

12. Hamer, D.H. and M. Walling, 1982. J. Mol. and Appl. Gen., in press.

13. Higgins, N.P. and N.R. Cozzarelli, 1979. In R. Wu, Ed., Methods in Enzymology, Vol. 68, Academic Press, New York, p. 50.

14. Jeffries, R. L., 1979. In D.W. Rains, R.C. Valentine and A. Hollaender, Eds., Basic Life Sciences, Vol. 14, Genetic Engineering of Osmoregulation, Plenum, New York–London, p. 135.

15. Kagi, J.H.R. and M. Nordberg, 1979. In Metallothionein, Birkhauser Verlag, Basel.

16. Kelley, D.B., J. D. Norlyn and E. Epstein, 1979. In J.R.
    Goodin and D.K. Northington, Eds., Arid Land Plant
    Resources, Texas Technical University Press, p. 326.

17. Kornberg, T. and M.L. Gefter, 1971. Proc. Nat. Acad. Sci.
    USA, 68:761.

18. Lautenberger, J.A., R.A. Schulz, C.F. Garon, P.N. Tsichlis
    and T.S. Papas, 1981. Proc. Nat. Acad. Sci. USA, 78:1518.

19. Lovett, P.S., and K.M. Keegins, 1979. In R. Wu, Ed., Methods
    in Enzymology, Vol. 68, Academic Press, New York, p. 342.

20. Mayo, K.E., R. Warren and R.D. Palmiter, 1982. Cell, 29:99.

21. Maxam, A., and W. Gilbert, 1977. Proc. Nat. Acad. Sci. USA,
    74:560.

22. Messing, J., R. Crea and P.H. Seeburg, 1981. Nucleic Acids
    Research, 9: 309.

23. Mizutani, S., and H.M. Temin, 1970. Nature, 226:1211.

24. Morrow, J.F., 1979. In R. Wu, Ed., Methods in Enzymology,
    Vol. 68, Academic Press, New York, p. 3.

25. Norlyn, J.D., 1979. In D.W. Rains, R.C. Valentine and A.
    Hollaender, Eds., Basic Life Sciences, Vol. 14, Genetic
    Engineering of Osmoregulation, Plenum, New York-London, p.
    293.

26. Otten, L., H. DeGreve, J.P. Hernalsteens, M. VanMontagu, O.
    Schieder, J. Straub and J. Schell, 1981. Mol. Gen. Genet.,
    183:209.

27. Pavlakis, G.N., N. Hizuka, P. Gordon, P. Seeburg and D.H.
    Hamer, 1981. Proc. Nat. Acad. Sci. USA, 78:7398.

28. Rick, C.M., 1972. In A.M. Srb, Ed., Genes, Enzymes and
    Populations, Plenum, New York, p. 255.

29. Roeder, R.G., 1976. In Ruhosick and M. Chamberlin, Eds.,
    RNA Polymerase, Cold Spring Harbor Lab, New York, p. 285.

30. Sanger, F., O. Nicklen, and A.R. Coubon, 1977. Proc. Nat.
    Acad. Sci. USA, 74:5463.

31. Schell, J. and M. VanMontagu, 1977. In A. Hollaender, Ed.,
    Genetic Engineering fro Nitrogen Fixation, Plenum Press, New
    York, NY, p. 159.

32. Simpson, R.T., 1981. In T. Papas and J. Chirikjian, Eds., Gene Amplification and Analysis, Vol. 2, Structural Analysis of Nucleic Acids, Elsevier-North Holland, New York, p. 347.

33. Suttcliffe, J.G., 1978. Nucleic Acids Research, 5:2721.

34. Vournakis, J., et. al., 1981. In T. Papas and J. Chirikjian, Eds., Gene Amplificaion and Analysis, Vol. 2, Structural Analysis of Nucleic Acids, Elsevier-North Holland, New York, p. 268.

35. Webb, M., Ed., 1979. In The Chemistry, Biochemistry and Biology of Cadmium, Elsevier-North Holland, New York.

36. Webb, M., and K. Cain, 1982. Biochem. Pharmac., 31:137.

37. Williams, B.C., and F.R. Blattner, 1980. In J.K. Setlow and A. Hollaender, Eds., Genetic Engineering, Plenum, New York, p. 201.

38. Zambryski, P., M. Holoters, K. Kruger, A. Depicker, J. Schell, M. VanMontagu and H.M. Goodman, 1980. Science, 209:1385.

Potentials in Aquaculture

Scott Sindelar

DeKalb AgResearch, Inc.
DeKalb, Illinois

Abstract

Increasing demand for seafood products and decreasing availability of raw materials from traditional sources have caused
seafood suppliers to look for an alternative supply source.
Aquaculture is the most feasible alternative and an attractive
business opportunity for several reasons, including lower production costs compared to other animal protein sources, emerging
new technologies, and an existing market infrastructure.

Aquaculture is the controlled, or partially controlled,
production of finfish, shellfish, or crustacea in fresh water;
mariculture is aquaculture's saltwater equivalent. These two
activities have been pursued for centuries in certain parts of
the world; however, it has only been recently that the significance of these pursuits has been more widely recognized. This
paper will discuss potentials in aquaculture (which will include
mariculture) from a corporate or business perspective. Potentials are here defined as opportunities to achieve economic
benefits.

In this paper, we will not consider fish for the manufacture
of fish meal. Currently, 31 percent of the total annual world
fish catch (about 73 million tons) is used for processing into
fish meal.

Need for a Future Alternative Source of Seafood
    The U.S. Department of Commerce, National Oceanic and
Atmospheric Administration (NOAA) estimates traditional world
stocks of wild seafood have a maximum harvest level of 100 to
150 million metric tons. Annual catches presently exceed 70
million tons, causing NOAA to state, "On a worldwide basis, a
shortage of fishery products can be expected within ten years if
the population continues to increase as forecast," (National
Oceanic and Atmospheric Administration, 1977). In the U.S. the
situation appears to be worse. Traditional fishery stocks are
being harvested near maximum sustainable yields, yet this amount
is not sufficient to supply domestic demand. The U.S. imported
a net 1.8 billion pounds of edible fish in 1979, more than one-
third of total U.S. fish consumption for a net value of $1.6
billion (National Marine Fisheries Service, 1979).
    Decreasing harvest and anticipated increasing world demand
for fish are expected to limit future availability of wild seafood. The National Marine Fisheries Service (NMFS) believes the
supply available from traditional stocks to satisfy domestic
demand will become inadequate during the 1980s, resulting in
shortages and increased prices (National Marine Fisheries
Service, 1975). This can already be seen in prices associated
with such products as clams, tuna, salmon, and shrimp (Table I).

Table I. Average annual price increase (ex-vessel)

Species	1967 to 1979	1975 to 1979
Haddock	9.6 percent	6.6 percent
Salmon (ave.)	16.3	16.3
Tuna (ave.)	10.5	13.5
Shrimp	13.3	19.6
Oysters	5.3	11.6
Ocean Perch	15.2	19.9
Lobster	8.4	6.3
Clams (soft)	13.6	16.3
Shellfish (other)	12.7	20.4
All Edible Fish	13.0	17.1

Aquaculture as an Alternative

Aquaculture has received attention as an alternative method
of increasing the supply of protein in the form of seafood.
World aquaculture production has doubled during the past five
years and now amounts to six million metric tons, or roughly ten
percent of total seafood production. NMFS indicates private
aquaculture firms in the U.S. currently account for the production of 40 percent of the nation's oysters, the majority of
catfish, crawfish, and trout and small quantities of several
other species for a total of 65,000 metric tons (National
Oceanic and Atmospheric Administration, 1977). However, this
still represents only three percent of domestic consumption.

Why Aquaculture?

The underlying potential of aquaculture arises from
economics of supply and demand. There are seven main reasons
why these economies make aquaculture a promising alternative.

   1. Increase in Domestic Consumption

Even though the U.S. is relatively low in world ranking of
per capita consumption of seafood (Table II), there is a dis-
cernible trend towards higher consumption since 1955. (Table
III). This upswing in demand has been caused by five factors
which should continue to hold in the future: the economic
advantage of seafood over certain other protein forms; continued
increase in dining out; increased health awareness; improved
product availability; and greater availability of fresh seafood.

Higher red meat prices will continue to cause consumers to
shop for different and less expensive protein sources on a cost
per serving basis. Seafood prices have increased during the
past decade, but not as rapidly as red meat prices. The
difference should become greater in the future due to technological advances and economies of scale in seafood production.
Technological advances, which will increase production per
dollar, are occurring in aquaculture, whereas no such advances
can currently be projected for meats and poultry. Economies of
scale will allow the reduction of distributors' mark-ups in
fishery products. At present, the ex-vessel, or ex-pond, prices
of seafood average only 39 percent of total retail value compared with 55 percent for beef and 43 percent for pork; the
remaining costs are distribution and marketing costs. As sales
increase and products become available on a regular basis, the
distribution and marketing percentage should fall, passing-along
savings to the consumer.

The continued increase in away-from-home dining and fast
food business should also increase consumption. The food
services trade accounted for 68 percent of retail sales of fish
and seafood products in 1979.

Seafood is perceived by society and the medical profession
as a healthy food. In many ways, fish is superior to meat and
poultry; while the protein content is comparable to meat, the

Table II. Per capita world fish consumption by Country

Country	Annual Per Capita Consumption Pounds Round Fish
Japan	151.7
Iceland	145.5
Norway	103.6
Denmark	76.6
Korea	75.4
France	47.4
United Kingdom	41.7
Canada	35.9
United States	34.6
Israel	33.1
Australia	31.1
Germany (West)	25.4
Nigeria	15.0
Mexico	10.1
India	6.4

fat and calorie contents are generally lower. This is important, as people have become more health conscious.

The better year-round availability of supply resulting from aquaculture will increase demand through increased marketing. Currently, food service businesses and retailers are reluctant to promote seafood products due to lack of a consistent supply. One of the reasons for increases in both catfish and trout consumption has been the industry's ability to overcome this problem, thus allowing retailers to advertise the product.

Consumer preference for fresh over frozen or canned products will also increase consumption because aquaculture can provide fresh seafood. Consumers would buy more seafood retail if they

Table III. Fish consumption in the U.S.

Period	Population (Million)	Fresh & Frozen	Canned	Cured	Total
1955–1959	169.1	5.7	4.1	.7	10.5
1960–1964	183.7	5.8	4.2	.5	10.6
1965–1969	195.3	6.1	4.3	.5	10.9
1970–1974	206.1	7.3	4.7	.4	12.2
1975–1978	213.9	7.9	4.5	.5	12.9

could see these products in the meat case consistently and be
assured they were truly fresh. Food service operators in a
number of interviews voiced the same rationale. They are
reluctant to add seafood to their product lines unless they can
be assured of a consistent fresh supply. Their customers
invariably ask, "Is it fresh?" In the white tablecloth trade,
this is particularly apparent.

2. Decrease in Traditional Supply Sources
The availability of traditional supplies of seafood for
domestic consumption is expected to continue to decline for the
following reasons: decrease in the number of fishermen, marine
pollution, overfishing of certain species, and less import
availability.

Fishing costs have escalated rapidly over the past few years
and should continue to do so. Diesel fuel and labor costs have
been the two primary forces. Diesel fuel prices have tripled
over the past few years, and there is no apparent immediate
method of improving the fuel efficiency of vessel operations.
Also, there are no forseeable technological advances which will
improve the efficiency of finding and capturing fish. Labor
costs have also escalated. Seldom do fishermen work at specified rates; rather, boat owners and crew share risk and rewards
by splitting proceeds. Due to union and general labor pressures
in some parts of the country, the crew's share has risen to as
much as 60 percent of the total.

As a result of these costs, the profitability of fishing has
decreased, causing fishermen to seek alternative employment. In
the Gulf Coast area, for example, oil operations offer more
attractive jobs than commercial fishing. In addition, some
shrimp and commercial fishing boats are being re-outfitted to
supply offshore oil drilling operations. Thus, the number of
market suppliers is shrinking.

Nearshore pollution threatens domestic supply. In 1975, NOAA estimated that the U.S. fish capture within three miles of its shore accounted for 45 percent of the total value of the U.S. catch. Fish and shellfish caught within three to twelve miles of the shore represented 15 percent of the total value (Frost and Sullivan 1977). These percentages have not changed drastically since 1975, thus nearshore pollution from industrial waste and sewage affects marine life approximating 60 percent of the income from the domestic catch. The effect is greater on shellfish than on finfish, which partially explains the rapid growth of cultured shellfish. Though government regulations have aided in cleaning up coastal water as well as commercial rivers and lakes, there are still serious problems. In certain waters, though clean now, one major spill or leak could deplete a seafood population for several years until it has an opportunity to rebuild itself.

Pollution has not only caused concern for dying species, but also for human health if seafood is contaminated. Thus, the government should act favorably toward seafood production in a controlled environment.

Overfishing in many cases has depleted U.S. coastal species. For example, abalone has been overfished off coastal California, making it the most expensive seafood item in America today, with menu prices often approaching $25 per serving.

Many techniques have been employed to prevent overfishing. Sea Grant and other governmental agencies have been trying in the past few years to cope with the problem. Undoubtedly, programs to sustain wild populations will follow. However, populations of many species may not be sustainable.

Increased worldwide consumption of seafood products is expected to continue. Many net seafood exporting countries are expected to retain a larger percentage of their seafood products to feed their own growing populations. Fewer imports will be available to the U.S., resulting in higher prices.

3. Lower Feed Requirements

The price of meat products is driven predominately by feed costs. The more efficiently an animal converts feed to meat, the greater its advantage in the face of rising feed prices. The average conversion rate for fish is 1.8 pounds of feed per pound of meat. This compares with 2.0 pounds for broilers, 2.5 pounds for eggs, 4.0 pounds for pork and 8.0 pounds for beef. Consequently, if feed costs went up $10 per ton, this would amount to a 4-cent production cost increase per pound of beef, compared to only an 8/10-cent similar increase per pound of fish.

The importance of feed conversion efficiency is shown in the following example. Assuming grain prices will increase at a rate equal to the rate of inflation, and assuming a 10 percent inflation rate over the next several years, feed prices should increase roughly $15 per ton annually. Thus, a $150 per ton

increase in feed prices would occur by 1988 and result in
increased production costs of 13 cents per pound of fish,
30 cents per pound of pork, and 60 cents per pound of beef.

On the other hand, some seafoods depend little if at all on
grain prices. Clams, oysters, scallops and mussels feed on
algae, which are either produced naturally or artificially in
tanks. Similarly the release-and-return salmon business is
relatively independent of feed prices. Salmon can be raised to
approximately 1/2 pound in confinement, then released for "ocean
ranching". Their weight on return averages 7 pounds -- for
which minimal feed costs have been incurred.

4. Technological Advances in Aquaculture

Neither the U.S. Department of Agriculture (USDA) nor the
Department of Interior (DOI) has in the past invested heavily in
aquaculture research because of perceptions that fish culture
was outside both jurisdictions; USDA considered fish culture a
means of producing only sport fish, and DOI perceived it as food
production. However, in the last few years, USDA has shown much
greater interest in aquaculture and has become more heavily
involved in information gathering as well as research in both
catfish and trout and in improving seafood marketing techniques.

There has also been a rapid rise in both research and educa-
tional programs in a number of universities including Oregon,
Washington, Wisconsin, California, Georgia, Auburn, Louisiana
State, MIT, Mississippi State, Arizona, Maine, New York and
Rhode Island. Private industry, too, has made significant
commitments to developing aquaculture. Thus, technological
advances in the production of seafood are expected in the near
future. One example is in the production of marine shrimp.
Until recently, attempts to get shrimp to reproduce in captivity
had been unsuccessful. Thus, all shrimp mariculture operations
had to be supplied with gravid females captured from the wild.
Now, however, several companies are successfully operating
shrimp hatcheries.

Another opportunity for technological change in the shrimp
industry lies in reducing feed costs. There has been little
competition among suppliers of shrimp feeds; currently feed
sells for as much as $500 per ton. Analysis by feed millers
indicates that feed of comparable quality could be manufactured
for 80 percent of that cost. Such a saving would reduce the
total production cost of green shrimp tails by 20 cents per
pound from the current $2.60 to $3.00 per pound level.

The greatest technological potential for aquaculture is
based in developing breeding techniques. Even in the most
studied species little work has been done to improve genetic
qualities. In a study done at DeKalb, one scientist estimates
that genetic improvements could decrease the cost of catfish
production by 3.6 cents per fish per year over a five-year
period, for a total of 18 cents per fish (DeKalb AgResearch,
Inc., 1979). If such genetic work had been done, catfish

producers could have saved $3.6 million this year. There are
numerous other opportunities for technological advance, all of
which should help promote the aquaculture industry.

5. Existing Marketing Infrastructure
The seafood industry has a marketing infrastructure already
in place. Product promotion is needed, but distribution and
brokerage channels have already been established. Thus, many
start-up costs associated with new ventures may be avoided in
the case of aquaculture.

6. Highly Fragmented Industry
The seafood production industry is highly fragmented. Most
participants are small operators who typically handle only one
product and who individually have a negligible effect on the
market. This creates an opportunity for a larger firm to step
in, earn a significant market share, become a price leader, and
utilize existing economies of scale.

7. Participants are Undercapitalized
Most operations in the aquaculture industry are undercapi-
talized. They cannot generate the funds needed to expand or to
implement new technology. Thus, sufficiently-funded firms are
able to enter the market and achieve significant advantages.

Aquaculture's Future
Taken together these seven factors provide aquaculture with
a bright future. The last three reasons, in particular, make
aquaculture attractive to large, amply-funded commercial
entities. In turn, such entities should help provide funds for
aquaculture's growth.
One consulting firm prepared a study in 1979 forecasting
that U.S. aquaculture production would become nearly a one-half
billion dollar market by 1990 (International Resource Development, Inc., 1979). A similar study, undertaken by another firm
in 1977, forecast the 1985 market at $300 million (Frost and
Sullivan, Inc. 1977). DeKalb's internal projections are shown
in Table II. This projected growth in aquaculture is based
primarily on a shift in supply rather than on increases in per
capita consumption. The supply of seafood, which currently
comes from wild sources, is expected to be derived more heavily
from aquaculture in the future.

Conclusion
The combination of increasing demand, increasing harvesting
costs, and a limited sustainable supply has created a need for
an alternative seafood supply source. Aquaculture is the best
alternative in meeting this need due to lower feed requirements
when compared to other animal protein sources, developing new
technologies, and an existing market infrastructure which

currently possesses a favorable investment climate. Aquaculture
should be viewed as a growth industry which has already
established many characteristics of success.

References

1. DeKalb AgResearch, Inc., 1979. Unpublished study, DeKalb, IL.

2. Frost & Sullivan, Inc., 1977. The fish farming market, New
   York, NY.

3. International Resource Development, Inc., 1979. Aquaculture
   in the 1980s, Norwalk, CT, p. 4.

4. National Marine Fisheries Service, 1975. National plan for
   marine fisheries, Washington, DC.

5. National Marine Fisheries Service, 1980. Fishery statistics
   of the United States, 1979. Current Fishery Statistics No.
   8000, Washington, DC.

6. National Oceanic and Atmospheric Administration, 1977. In
   John Glude, Ed., NOAA Aquaculture Plan, Washington, DC, p. 1.

Aquabusiness

Harold H. Webber

Groton BioIndustries
Groton, Massachusetts

Abstract

The sequence of decisions that must be made by corporate and
national planners before committing an investment in marine
aquaculture is considered here. Certain criteria for the
selection of species to be cultured, the technologies to be
employed, the selection of sites, some problems in transition
from experimental methods to commercial scale operations and the
nature of the investment considerations are discussed.

Recent technological advances in mariculture have been made
as a consequence of intensified research in marine biological
sciences during the period since World War II. These have
generated sufficient new baseline data regarding behavior,
reproductive physiology, nutrition, disease, parasite controls,
and the various ecological interactions which now make it
possible to confine high population densities in ponds and cages
and in raceways in which the water is flowed through or recycled

Vertically Integrated Aquaculture

The rate of change in technology has been increasing in
recent years, as more public and private research and development funding has been applied, and as the need and prospects for
commercialization of mariculture have become more widely
recognized. This is a consequence, in part, of the increasing
costs incurred in the capture fishery, while the unit cost of

115

the culture fishery promises to be reduced, or remain level, as
productivity increases. However, results of recent research and
development in the biological sciences have not yet been integrated into the broader context of large scale, vertically
integrated, high technology, centrally controlled, aquabusiness
food production systems. We shall call such operations
Vertically Integrated Aquaculture (VIA).

These production systems require the integration of a very
large number of variables, some of which are uncontrollable but
may be predictable, and some of which, although controllable,
are not yet well enough understood and need further elucidation
in terms of systems engineering criteria. Much of the baseline
data generated in the laboratory have not yet been translated
through critical scale-up trials into commercial operations.
Often, the prudent, cautious, scientific experimentalist has
been enticed by highly venturesome and impatient entrepreneurs
into premature full-scale application of concepts and techniques
only recently validated at the laboratory level. Consequently,
many mariculture ventures have been destined to failure because
of the unwise and hasty judgments of investors whose appreci-
ation of the complex technology was often as faulty as their
business judgment.

Mariculture, like most of the bioindustries which require
critical ecosystems management, is extremely site-sensitive,
highly dependent upon specific characteristics of the ecosystem
and the society in which it is to be undertaken. Although
mariculture has long been practiced as an art, tailored to the
site and the socioeconomic context, it is only recently that we
have gained sufficient biological and bioengineering knowledge
and an understanding of the social and institutional influences
on biobusiness to enable us to translate the arts of mariculture
-- by quantification, integration and generalization -- into
practical commercial technologies.

We are finding that these technologies may, in fact, be
transferable from the ecosystems in which they are generated to
others around the world. The controlling word here is 'around',
by which I mean within latitudinal limits around the globe,
where ecosystem characteristics are sufficiently similar to
enable technology transfer with little adaptaton to accommodate
site variability. However, the many social, cultural/
institutional, and economic characteristics of the various
societies that share a latitude must be understood and
incorporated into the plan for a mariculture venture.

Whereas similarities among ecosystems that share latitudes
should be self-evident to a mariculturist transferring
technologies, dissimilarities along the meridians should be
equally evident. Yet much effort and money has been spent in
attempts to shift mariculture techniques from the temperate
zone, where they were developed, to tropical coastal zones.

I will now consider the essential components of marine
aquaculture production systems, which are designed to fulfill

the goals of national and/or corporate planners whose investment
criteria are social (job creation), political (foreign exchange
improvement and infrastructure development), or solely economic
(profits). There exists, however, considerable uncertainty
among planners regarding the degree of risk and the extent of
the rewards -- the investability -- that marine aquaculture
entails.

Structuring an Aquabusiness Plan

For traditional production of aquafoods by subsistence
farming in fresh, brackish and even marine waters, yields of
200-500 kilograms per hectare were considered to be normal; such
harvest ratios reflect the natural carrying capacity of the crop
animal in most coastal ecosystems in the tropics. Seeding of
the farm was usually accomplished by natural recruitment, or
gathering the young in the wild. However, large scale industrial mariculture, demanding substantial capital investment and
operating funds in order to yield commensurately high profits,
requires much higher bioproductivity. This can best be achieved
by providing high stocking densities by seeding with hatchery
generated juveniles. Crop population densities beyond the
natural carrying capacities demand energy subsidies in the form
of feed, aeration, water exchange, and in some instances, heat.
The maintenance of high water quality in such high intensity
systems requires an unusual degree of monitoring and understanding of the water condition and the chemistry of saline waters,
and a highly responsive controlling mechanism to ensure a
salubrious environment for growth and survival. Fast remedial
action is also required for disease/parasite and predator/
competitor problems. As a consequence of these unnatural
conditions, a higher level of technical sophistication is needed
for the management of the amended aquacultural ecosystems than
was necessary for the subsistence farming operations in the
natural ecosystems of traditional coastal zone mariculture.

The complexity of the production system, resulting from the
high productivity technologies and the diversification of
operations from spawning in a hatchery through larviculture,
nursery, and grow out in the production sector to harvesting,
processing, packaging, and distribution in the market sectors
require very special management skills uncommon in most
industries.

Furthermore, VIAs are generally capital-intensive enterprises and require three to five years to reach the stage of
positive cash flow. Thus, they demand an in-depth capital
resource that can sustain several years of losses normally
expected during start-up, as well as inordinate, unexpected
perturbations to which high risk bioindustries are frequently
subjected. Many failures in aquabusinesses have been due to
undercapitalization.

Market Considerations and Species Selection

Since the primary motivation for mariculture ventures is to supply food for human consumption, the design of the enterprise and the management rationale should be guided by market parameters. Probably the first among the criteria that will measure the marketability of aquafoods are the traditional food habits and taboos in the different marketplaces. From among the thousands of aquatic plants and animals that inhabit the fresh and saline waters of the earth, only a certain few preferred species are being investigated as candidates for domestication and culture. These species are chosen because of high consumer acceptance which contributes to high market values and justifies the risk to the investor. Among the reasons for consumer acceptance of these species is that they have been sufficiently abundant to warrant capture, and, therefore, they are familiar aquafoods. Ease of preservation and preparation and the degree to which the user's basic nutritional requirements can be satisfied are also important criteria. In addition, social identity with certain foods leads to status-seeking in expressing food preferences. Such subjectively judged characteristics as flavor, texture, and certain appearance factors (color, luster, and form) contribute significantly to consumer preferences and acceptance of aquafoods. Hedonic testing clearly reveals that heavy reliance on these subjective factors by the culturist is valid; since aquacultural conditions can significantly alter traditional product characteristics, the culturist should be alert to organoleptic changes, such as the appearance of off-flavors or soft textures that may affect the market acceptance of cultured aquafoods.

Other market criteria which must be considered relate to problems of manufacturing efficiency, such as dressout percentages. Ultimately, the essential market criteria that will influence the choice of candidates for culture must be the size and structure of the market and the elasticities of demand and price for the aquafood products that can be grown.

Technology

There has been a tacit assumption of an analogy between terrestrial and aquatic animal husbandry practices. However, to the extent that the analogy is true, it is much more applicable to freshwater farming, such as for carp, catfish, and tilapia, than it is for the marine species which are to be managed in more hostile environments. Salt-induced corrosion, the interaction of the crop animal with predators and competitors, and the unwise location on coasts subject to high energy wave action, storm, and erosion make comparison with more benign conditions for poultry, swine, and cattle production less tenable. Furthermore, a culture technology has high risk unless control over the reproductive cycle of the crop species is guaranteed so as to ensure an adequate seed supply for stocking

the next generation. This has been more readily achieved with
freshwater finfish than with marine fish, due in large part to
our lack of knowledge of fish behavior, reproductive physiology,
diseases, and basic nutritional requirements. The problems of
feeding small-mouth fish larvae and achieving high survival in
larviculture of marine species have been major constraints on
commercial mariculture development.

As seed production problems are ameliorated, as they have
been in some flatfish, the next limiting factor is nutrition
during the growout phases. Since the cost of feed and feeding
represents the single largest item in the operating budget, any
gains in efficiency through improved nutrition and feeding
technology resulting in greater growth and survival can make a
significant contribution to profitability. When high density
populations are confined in ponds, cages, raceways, or otherwise, a feed supplement in some cases, or a complete ration in
others, must be provided to ensure the growth, health, and
survival of the crop. Such rations are usually formulated from
agricultural residues, oilseed cakes, various food processing
wastes, and most important of all, fishmeal, which not only
supplies essential amino acids, but evidently growth promoting
substances and attractants as well.

Very little classical nutrition research has yet been
performed with the candidates for marine aquaculture, and until
we are better informed regarding the essential components and
proportions of diets, we will not optimize the system for high
profitability.

The lack of control of diseases and parasites in hatchery
operations and in nursery/growout continues to limit large scale
operations. Occasionally predation and poaching have been
limiting, and in some environments institutional constraints
have made the conduct of a business difficult and costly.

Locating the Aquafarm

Following the evaluation and selection of the candidate
species appropriate to satisfy a given market and the determination of the most suitable culture technology, one must then
make a most critical judgment concerning site selection. In
this regard, satisfying the ecological requirements, such as
water quality, soils, meteorological criteria, etc. may not be
sufficient in itself; for, as noted, the economic and political
climate and cultural and sociological milieu can also be
limiting influences on the successful application of a business
plan that is otherwise well based on marketing and technological
judgments.

Poor site selection has in many cases in the United States
accounted for the collapse of an enterprise. Hurricanes, water
quality and supply, pollution, regulatory constraints, or
opposition to the venture by the indigenous fishing community
have been site-related causes of failure.

Transition from Laboratory to Commercial Scale

Little information has been published about the design, construction, and operation of large scale, commercial marine aquaculture systems. This is generally judged to be proprietary information, but most of it is based on unsuccessful ventures. VIA ventures are commercial enterprises where all inputs and outputs including labor, land, and product have market values, and the underlying motivation is to produce, at a profit, substantial volumes of aquafoods for sale at the marketplace.

Scale-up of experimental installations to larger size is not for size, per se, but is required to achieve the volume of sales sufficient to justify the high investment cost of research, development and start-up, and to gain the advantages of economies of scale. The nature of the development process and the magnitude of the problems involved in the transition from laboratory development and testing to pilot plant to full-scale operations have not often been understood in light of the complexity of the systems, particularly in vertically integrated operations. Management decision criteria undergo dramatic change as design parameters change with scale.

Efforts to optimize integration of diverse and interacting variables as well as the shift in emphasis from laboratory precision to production operations with satisfactory quality control criteria, demand the experienced management of complex biological systems which has yet to find its way into the conduct of large, centrally controlled VIAs.

There is an urgent requirement for a pilot farm to be operated as part of an ordered sequence of development steps in order to adapt the extremely site-sensitive production technology to the specific environmental conditions in which it must operate. However, a pilot farm is rarely required to determine whether to undertake the venture. Generally, this decision can be arrived at after a pre-investment feasibility study and a detailed analysis of the business plan have been worked out.

With appropriate experiments to evaluate the interaction of the many influences on growth and survival of animals in high intensity culture systems, a pilot farm can provide valuable day-to-day, as well as long range, guidance on how to manage the system. The interactions of such controlled variables as stocking rate, ration composition, pond fertilization, feeding amounts and frequencies, water exchange rate and/or aeration, etc., must be well understood and optimized before the goals for high productivity can be realized.

Investability

The decision to invest significantly in aquafarming, whether it be by a private entrepreneur or by a public administrator, should be predicated on an in-depth analysis of the size and nature of the market and a determination of elasticities of price and demand of the particular species that are being

considered. Attention should also be given to the likely need
for product and market development programs. At the same time,
the state-of-the-art of the technological components must be
evaluated in order to determine the degree of risk being assumed
and the site selection criteria which must be satisfied to
ensure that a profitable system is feasible if the appropriate
management components are inserted.

Development and/or commercial bank financing are more likely
to be attainable following the completion of at least two crop
cycles in the pilot facility. With the experience and data from
a pilot operation, the investors will be more receptive to a
proposal from the entrepreneurs and thus provide a more favorable debt/equity ratio to enable early expansion to full scale.

Justification for investment in mariculture is measured by
the potential for sales revenue and profitability in the private
sector, and by rural employment and the generation of wealth in
the form of high protein foods in the public sector. Growth in
this bioindustry will continue to be realized only as corporate
and national planners gain the basic knowledge and the human and
financial resources required to achieve greater productivity and
appropriate scale for VIA.

Biotechnology: A Potential U.S. Contribution to Mariculture

John H. Ryther

University of Florida
Gainesville, Florida

Abstract

While technological advances have been made in mariculture
in the United States, sociological, legal, economic, and geographical obstacles have constrained the industrial development
of mariculture in this country. These constraints have encour-
aged the pursuit of mariculture work in the Third World where
regulatory and other problems are less severe. Introduction of
Western biotechnology to developing countries can improve stocks
of species suitable for mariculture there and reduce the cost of
growing them. With the exception of salmonids, oysters, and
some seaweeds, marine oganisms now reared in captivity are still
in their native, unimproved state. There is great potential for
genetic improvement, by classical and modern methods, of these
organisms, which are better suited in many ways to genetic
manipulation than terrestrial species.

Mariculture has not been conspicuous for its success in the
United States since its beginning two or more decades ago. Most
agree that the problems with the struggling new field are less
of a technical nature, but rather a combination of sociological,
legal and economic problems together with environmental concerns
and geographical limitations. Many marine organisms can now be
grown successfully in captivity throughout their life cycle, but
few can be reared and sold at profit.

This rather unsatisfactory situation has led aquaculturists, entrepreneurs, and venture capitalists who wish to invest in maricultural pursuits to look elsewhere, primarily in the tropical to semi-tropical maritime countries of the Third World, where relatively cheap labor, availability of coastal lands, and lack of significant reduction of the other non-technical constraints referred to above make such investment considerably more attractive.

In the developing countries, however, it is often technical problems that constrain the expansion of mariculture. For example, lack of hatchery technology for such important species as milkfish, mullet, penaeid shrimp, oysters, and other mollusks has resulted in a dependence for their culture upon the collection of juvenile animals from wild populations, a situation that has severely held back the growth and reduced the profitability of those otherwise highly successful culture practices. Lack of knowledge of the nutritional requirements of those and other important food species and of the technology of formulating and preparing complete-diet artificial feeds has resulted in dependence upon natural foods, which can be scarce, undependable, and prohibitively expensive when in short supply.

These technological advances have been and are now being achieved in the United States, Europe, and Japan, and are now in the process of being successfully introduced to the developing countries, particularly when accompanied by Western capitalization and entrepreneurial investment.

Potential Applications of Biotechnology

What of the modern biotechnology? How can it be applied to the field of mariculture and how can it be exported to the Third World? Undoubtedly, genetic engineering and recombinant DNA techniques may eventually be applied to the creation of new and improved strains of many if not all of the existing and potential mariculture species -- varieties that grow faster and larger, contain more meat and less waste, can grow within an expanded range of temperatures and salinities, are disease resistant, have behavioral characteristics amenable to mass cultivation (eg. lack of territoriality and cannibalism), and are more efficient in converting food to flesh. The list of potential improvements is long indeed.

What is the time scale for these advances? The new biotechnology is still in its very early stages of development. Production of eukaryotic proteins such as insulin and interferon from a few well characterized microorganisms has been the first industrial target, already close to realization. But application of genetic engineering techniques to the higher plants and animals is a long-term prospect, perhaps a decade or more away. What, in the meantime, can existing and available biotechnology do for the struggling new field of mariculture? I suggest that much, indeed, can be done, and the time is ripe for such action today. Further, I suggest that new and expanded programs and

facilities for improvement of mariculture species stocks, using
the existing biotechnology, will enhance and accelerate the
development of biotechnology in its application to mariculture
and will, in fact, provide a setting essential to the success of
the succeeding generation of biotechnological research.

Terrestrial plants and animals of importance to humans have
been domesticated for thousands of years, during most of which
time they have been subjected to at least a crude form of
selective breeding and stock improvement. During the past
century, following the discoveries of classical genetics, truly
astounding progress has been made in improved growth, yield,
disease resistance, range extension, and other desirable
properties of domesticated plants and animals.

Marine organisms that are now reared in captivity, on the
other hand, are for the most part still in their native,
unimproved form. The only exceptions are salmonid fishes,
oysters, and certain seaweeds, each of which will be discussed
later. The reason for this is, of course, that mariculture is
largely a new human endeavor, undertaken in a serious way only
within the past two decades.

During the early years, sea farmers were preoccupied with
problems of maturation, spawning, larval rearing, nutrition, and
prevention and treatment of diseases in their cultivated animals
and with the associated engineering problems of systems design
and operation. There was little time to worry about stock
improvement while the problems of growing existing stocks
remained unresolved. Accompanying these early efforts was a
concentration on very few species, namely those that were the
best known, most predictable and reliable, and easiest to grow.

Gradually, that situation has changed. Today dozens of species
of marine algae, invertebrates, and finfishes can be grown
throughout their life cycle in captivity. Experienced breeders
speak confidently of their ability to spawn and rear through
successive generations any marine finfish deserving of the
effort.

The principal problem with mariculture today, as mentioned
previously, is not the technological one of growing the organism, but the economic one of growing it at profit. But the two
are, of course, related. Improving the technology will normally
bring down the cost. This is precisely where genetic improvement of the raw material comes into play.

Advantages Over Terrestrial Species

In a thought-provoking review article, Wilkins (1981) points out
the several attributes that make aquatic organisms particularly
suited to genetic manipulation, thereby giving them a decided
advantage over terrestrial species. These are (quoting Wilkins
verbatim):

*(1) Domesticated livestock are homeothermic verte-
      brates in which levels of genetic variability,
      measured in terms of heterozygosity (0.05-0.06),
      proportion of polymorphic loci (0.10-0.20) and
      numbers of alleles per polymorphic locus (approx-
      imately 2) is low; cultivable aquatic animals are
      invertebrates (mollusca, crustacea, echinoder-
      mata) or poikilothermic vertebrates in which
      genetic variability, measured in the same way, is
      significantly higher. In molluscs, for example,
      heterozygosity is approximately 0.15, the pro-
      portion of polymorphic loci is approximately 0.35
      and an average of 3.9 alleles are segregating at
      each polymorphic locus. In other words, the
      amount of variation which is available for
      manipulation is greater in aquatic organisms.*

*(2) Domesticated livestock are already considerably
      improved by a long history of artificial selec-
      tion and their potential for further improvement
      may be limited. (Mayr 1963; Manwell and Baker
      1970). Aquatic organisms are still largely
      unaltered by artificial selection and their
      genetic repertoire reflects the action of natural
      selection in natural environments. The stabili-
      zation of environmental variation and the
      improvement in husbandry which accompany the
      domestication process add a directional element
      to selection such that different wild genotypes
      can be selected to suit different more stabilized
      artificial environments.*

*(3) Fecundity in fishes and many cultivable inverte-
      brates is very considerably greater than in any
      terrestrial vertebrates. Hence many more
      siblings can be raised from outstanding indi-
      vidual crossings, selection can be more intense,
      full sib and half-sib families each with
      statistically large numbers of offspring are
      easily produced, and larger numbers of broodstock
      with desirable characteristics can be reproduced.*

*(4) Sex determination is more plastic in fish and
      aquatic invertebrates, and sex-reversal and
      protandrous hermaphroditism are common features
      of many such organisms. Production of mono-sex
      stocks through hormonal treatment is easier, and
      self fertilization of hermaphrodites or of
      sex-changed females with sperm stored from their
      male phase, accelerates the production of fully
      inbred lines.*

*(5) Interspecies crosses are very often viable and
      even fertile in fishes and invertebrates.*

> *External fertilization and larval development*
> *make artificial interspecies mating easier to*
> *perform than for example in placental mammals.*
> These features suffice to indicate that genetic manip-
> ulation of existing variability in aquatic organisms
> may be easier to achieve than might at first appear,
> and genetic strategies unsuitable or impossible with
> domesticated livestock may be feasible or even
> desirable in aquaculture. The aquatic mode of life
> certainly imposes many restraints on the holding and
> rearing of these organisms but from a genetic
> viewpoint these are more than compensated by some of
> the genetic and reproductive features of their
> life-cycle.

And, in the summary of his article:

> Now that the technology of production and rearing has
> been largely mastered for many aquatic organisms it is
> reasonable to look to genetic techniques to contribute
> to the further improvement of cultivated stocks. Not
> only are the traditional techniques of selective
> breeding, inbreeding and interstrain crossing avail-
> able for this, but other procedures not possible or
> readily feasible with higher vertebrates are also
> possible. These include polyploidization, gyno-
> genesis, mono-sex culture and self-fertilization.
> These procedures are made possible largely by the
> greater flexibility inherent in external fertilization
> and by their very high natural fecundity. In these
> features aquatic animals resemble plants more than
> terrestrial vertebrates. The scope for genetic
> improvement conferred on aquatic animals by these very
> features exceeds that of domestic livestock and it
> will be surprising if, in the long term, novel
> improvements are not attained through the exploitation
> of these novel reproductive traits.

Wilkins did not address the potential for genetic improvement in aquatic plants, but there again the marine algae are
particularly suited to genetic manipulation. Many of the seaweeds of commercial importance to humans are characterized by a
complex alternation of haploid, sexual plants and diploid,
asexual sporophyte generations. Chinese culturists have found
the kelp, Laminaria japonica (an important species for food,
algin, and other uses in Eastern countries), to be highly
variable in several important characteristics (length, frond
width, growth rate, iodine content, temperature tolerance,
etc.), and that most of these features are controlled by single
gene or quantitative polygene inheritance patterns. (Fang et
al. 1962a, 1962b, 1963, 1965, 1966). Significantly, increased

yields have resulted from selective breeding of these variants and the resulting improved growth characteristics of the new strains of kelp, and through range extension by new warm-water-tolerant varieties to areas much further south than the species normally occurs.

More recently, Fang et al. (1977) have been able to propagate the microscopic female gametophyle of L. japonica in tissue culture, cause the growth of callus tissue which may be maintained as pure gametophyte clones, and induce the partheno-genic differentiation and growth of sporophyte plants from the clonal tissue cultures, thereby opening the door to pure culture propagation of the seaweeds in much the same way as is now done with higher terrestrial plants.

Some of the red algae (Rhodophyta) are even more susceptible to both natural genetic variability and manipulation, for the haploid gametophytes (male and female) are large, macroscopic seaweed plants that are in fact indistinguishable from the diploid, asexual sporophytes. The gametophytes, being mono-ploid, express all mutations phenotypically, sometimes in a bewildering variety of colors, shapes, and morphologies. J.P. van der Meer of the Atlantic Research Laboratory (Halifax, Novia Scotia) has published a series of papers on the genetics of Gracilaria tikvahiae (van der Meer 1977, 1978, 1979a, 1979b, 1981; van der Meer and Bird 1977; van der Meer and Todd 1977), that well illustrates the fascinating genetic plasticity of the species which spontaneously produces many mutants and, after metagenesis with ethyl methane-sulphonate, literally hundreds more (most, though not all) showing recessive, single-gene transmission characteristics. The sporophyte plants also release fertile diploid gametes from the sporophyte plants, combinations of which may produce triploid and tetraploid plants. Others (Cheney, this volume) are developing tissue culture techniques for G. tikvahiae with the object of attempting somatic hybridization of the species.

No one has yet screened the many natural and human-made varieties of this interesting alga for products of commercial value, though it contains the valuable polysaccharide, agar, which, if contained in mutant strains in higher quantity or quality than in the normal parent stock, could be of economic interest. There is as yet, however, no commercial mariculture industry for seaweeds in the Western world and, hence, no research effort directed specifically towards stock improvement of mariculture species, as is the case in China.

The oyster, on the other hand, has been cultivated in Europe since Roman times and is at present the only truly marine species that is successfully grown, from an economic viewpoint, in the United States. The hatchery production of seed oysters, complex alternation of haploid, sexual plants and diploid, asexual sporophyte generations. Chinese culturists have found the kelp, Laminaria japonica (an important species for food, algin, and other uses in Eastern countries), to be highly

variable in several important characteristics (length, frond
width, growth rate, iodine content, temperature tolerance,
etc.), and that most of these features are controlled by single
gene or quantitative polygene inheritance patterns. (Fang et
al. 1962a, 1962b, 1963, 1965, 1966). Significantly, increased
yields have resulted from selective breeding of these variants
and the resulting improved growth characteristics of the new
strains of kelp, and through range extension by new warm-water-
tolerant varieties to areas much further south than the species
normally occurs.

More recently, Fang et al. (1977) have been able to
propagate the microscopic female gametophyle of L. japonica in
tissue culture, cause the growth of callus tissue which may be
maintained as pure gametophyte clones, and induce the partheno-
genic differentiation and growth of sporophyte plants from the
clonal tissue cultures, thereby opening the door to pure culture
propagation of the seaweeds in much the same way as is now done
with higher terrestrial plants.

Some of the red algae (Rhodophyta) are even more susceptible
to both natural genetic variability and manipulation, for the
haploid gametophytes (male and female) are large, macroscopic
seaweed plants that are in fact indistinguishable from the
diploid, asexual sporophytes. The gametophytes, being mono-
ploid, express all mutations phenotypically, sometimes in a
bewildering variety of colors, shapes, and morphologies. J.P.
van der Meer of the Atlantic Research Laboratory (Halifax, Novia
Scotia) has published a series of papers on the genetics of
Gracilaria tikvahiae (van der Meer 1977, 1978, 1979a, 1979b,
1981; van der Meer and Bird 1977; van der Meer and Todd 1977),
that well illustrates the fascinating genetic plasticity of the
species which spontaneously produces many mutants and, after
metagenesis with ethyl methane-sulphonate, literally hundreds
more (most, though not all, showing recessive, single-gene
transmission characteristics). The sporophyte plants also
release fertile diploid gametes from the sporophyte plants,
combinations of which may produce triploid and tetraploid
plants. Others (Cheney, this volume) are developing tissue
culture techniques for G. tikvahiae with the object of
attempting somatic hybridization of the species.

No one has yet screened the many natural and human-made
varieties of this interesting alga for products of commercial
value, though it contains the valuable polysaccharide, agar,
which, if contained in mutant strains in higher quantity or
quality than in the normal parent stock, could be of economic
interest. There is as yet, however, no commercial mariculture
industry for seaweeds in the Western world and, hence, no
research effort directed specifically towards stock improvement
of mariculture species, as is the case in China.

The oyster, on the other hand, has been cultivated in Europe
since Roman times and is at present the only truly marine
species that is successfully grown, from an economic viewpoint,

in the United States. The hatchery production of seed oysters,
including the more recent production of "culchless" or
unattached seed, is also one of the few technological advances
in mariculture that has taken place in the United States. It is
perhaps not surprising, then, that the oyster is the one truly
marine animal to have been the subject of any significant
genetic research. The subject has been reviewed by Newkirk
(1980) and need not be discussed here in detail.

   In general, the research effort in molluscan genetics has
been diffuse, sporadic, and rather cursory. To date, the only
application of such research has been in the selective breeding
of strains resistant to disease (Haskin and Ford 1978) or environmental stress (Beattie et al. 1978). Nowhere is the kind of
large-scale, long-range program in place that would be needed
for a serious effort in molluscan stock improvement. Nowhere do
the facilities or the support for such a program exist.

Selective Breeding in Salmon

   The most successful application of classical selective
breeding techniques to mariculture has been in the improvement
of salmonid fish stocks. The pioneer in this field, Lauren R.
Donaldson of the College of Fisheries, University of Washington,
after devoting 38 years of selective breeding to the ultimate
production of his "supertrout", a superior strain of rainbow
trout, turned his attention to the Pacific salmon. Following
establishment of successful runs of chinook salmon in a small
stream that flows into Union Lake on the University campus,
Donaldson and his co-workers turned their attention to selection
for better growth and larger size, fecundity, time of return,
environmental tolerance, and disease resistance.

   Since his retirement, Donaldson's work has been carried on
by other State and Federal fishery biologists with continued
success. Now, after some 25 years of selective breeding,
returns of hatchery-reared smolt to the fishery, the bottom line
in such ocean ranching practices have increased from somewhere
in the order of 0.1 percent to 1 to 2 percent and now represent
a major contribution (40 to 50 percent) to the salmon fishery
(Ryther 1981).

   Similar selective breeding efforts are now underway with the
Atlantic salmon in the Canadian Maritimes, sponsored by provin-
cial government and private organizations. Norwegian biologists
are also engaged in various kinds of genetic research to improve
growth, control disease, and prevent sexual maturation in
Atlantic salmon used in their highly successful cage culture
industry.

   The markedly increased returns of Pacific salmon are only
partially attributable to the effects of selective breeding.
Improved nutrition and other factors have also contributed to
the hatchery production of larger, hardier smolts with greater
survivability. But genetic improvement of the stocks has been
an important element in the salmon success story that can be

pointed to as an example of the improved yields in mariculture that can be expected from the application of very simple, classical principles of genetics. Perhaps more important, it illustrates that the successful application of such principles is a long, slow process. Hopefully, modern biotechnology will accelerate the rate of progress. But it seems clear that a more general application of genetic principles and techniques to mariculture is long overdue, and that both classical and modern approaches, perhaps in concert, will prove profitable to a field that clearly needs the benefit of improved biotechnology.

References

1. Beattie, J.H., W. K. Hershberger, K.K. Chew, C. Mahnken, E.F. Prentice and C. Jones, 1978. Breeding for resistance to summertime mortality in the Pacific oyster (Crassostrea gigas), Washington Sea Grant Report WSG 78-8.

2. Fang, T.C., C.Y. Wu, and J.J. Li, 1962a. Increased adaptability to high temperature of gametophytes and sporlings of the Haiqing No. 1 breed of Laminaria japonica Aresch., Oceanol. et Limnol. Sinica, 4:29-37.

3. Fang, T.C., C.Y. Wu, B.Y. Jiang, J.J. Li and K.Z. Ron, 1962b. The breeding of a new breed of Haidai (Laminaria japonica Aresch.) and its preliminary genetic analysis, Acta. Bot. Sin., 4:197-209.

4. Fang, T.C., C.Y. Wu, B.Y. Jiang, J.J. Li, and K.Z. Ron, 1963. The breeding of a new variety of Haidai (Laminaria japonica Aresch.), Scientia Sinica, 12:1011-1018.

5. Fang, T.C., B.Y. Jiang and J.J. Li, 1965. Further studies on the genetics of Laminaria frond-length, Oceanol. et Limnol. Sinica, 7:59-66.

6. Fang, T.C., B.Y. Jiang and J.J. Li, 1966. The breeding of a long-frond variety of Laminaria japonica Aresch., Oceanol. et Limnol. Sinica, 8:43-50.

7. Fang, T.C., C.H. Tai, Y.L. Ou, C.C. Tsuei and T.C. Chen, 1978. Some genetic observations on the monoploid breeding of Laminaria japonica, Scientia Sinica, 21:401-408.

8. Haskin, H.H. and S.E. Ford, 1978. Mortality patterns and disease resistance in Delaware Bay Oysters, Proc. Nat. Shellfish Assoc., 58:80-86.

9. Newkirk, G.F., 1980. Review of the genetics and the potential for selective breeding of commercially important bivalves, Aquaculture, 19:209-228.

10. Ryther, J.H., 1981. Mariculture, ocean ranching and other culture-based fisheries, Bioscience, 31:223-230.

11. van der Meer, J.P., 1977. Genetics of Gracilaria sp. (Rhodophyceae, Gigartinales). II. The life history and genetic implications of cytokinetic failure during tetraspore formation, Phycologia, 16:367-371.

12. van der Meer, J.P., 1978. Genetics of Gracilaria sp. (Rhodophyceae, Gigartinales). III. Non-mendelian gene transmission, Phycologia, 17:314-318.

13. van der Meer, J.P., 1979a. Genetics of Gracilaria sp. (Rhodophyseae, Gigartinales). V. Isolation and characterization of mutant strains, Phycologia, 18:47-54.

14. van der Meer, J.P., 1979b. Genetics of Gracilaria tikvahiae (Rhodophyceae). VI. Complementation and linkage analysis of pigmentation mutants, Can. J. Bot., 57:64-68.

15. van der Meer, J.P., 1981. Genetics of Gracilaria tikvahiae (Rhodophyceae). VII. Further observations on mitotic recombination and the construction of polyplids, Can. J. Bot., 59:787-792.

16. van der Meer, J.P. and N.L. Bird, 1977. Genetics of Gracilaria sp. (Rhodophyceae, Gigartinales). I. Mendelian inheritance of two spontaneous green variants, Phycologia, 16:159-161.

17. van der Meer, J.P. and E.R. Todd, 1977. Genetics of Gracilaria sp. (Rhodophyceae, Gigartinales). IV. Mitotic recombination and its relationship to mixed phases in the life history, Can. J. Bot., 55:2810-2817.

18. Wilkins, N.P., 1981. The rationale and relevance of genetics in aquaculture: an overview, Aquaculture, 22:209-228.

II. MARINE PARMACEUTICALS AND BIOPRODUCTS

Organic Chemicals from Marine Sources

George Whitesides

Harvard University

Janet Elliott

BioInformation Associates

Abstract

Research opportunities for organic chemicals from the sea
can be divided into three classes: near-term (less than 5 yrs
to fruition), more distant, and very speculative, long-term
opportunities. This paper describes some of these opportunities
and gives the authors' view on their possibilities for success.
Near-term projects might include development of by-products from
the fish and macroalgae industries, process improvements, development of new materials for aquaculture, and the development of
new products such as unusual sugars, polysaccharides, carotenoids, and algal lipids. More distant opportunities could
include energy farming, pharmaceuticals, polysaccharides to be
used for oil production and transport, chemicals to control
fouling, and agricultural chemistry for marine systems. Long-
term opportunities are more speculative and include increased
exploitation of marine organisms as representing a tremendous
untapped gene pool, using or mimicking the ability of many
marine organisms to selectively extract useful elements from
seawater, using bacteria from rift geothermal sources to utilize
H_2S, exploiting unique algal photosynthetic systems, and
controlling the movements of marine animals during harvesting by
using pheromones and communication systems.

This conference has summarized some of the interesting
characteristics of marine organisms as sources of food and
chemicals. The predominate point of view has been that of
enthusiasts who are knowledgeable in marine biology. This paper

will cover some material which has been presented previously. It differs from previous presentations in starting from a chemist's perspective. The authors wish, in particular, to inject several practical questions into these discussions:

(1) What products which might be obtained from marine sources are actually needed, and at what price can one afford to make them?

(2) Can marine biotechnology compete with more conventional areas of biotechnology for increasingly scarce research funds? For example, it appears likely that, given sufficient time and effort, genetic engineering of algae could become a reality. It is even possible that genetically engineered algae could be found which would flourish in the wild. Are the best opportunities for research in biotechnology to be found in algae, or in working with industrially useful organisms such as the yeast <u>Saccharomyces</u> or the bacterium <u>Bacillus</u> <u>subtilis</u>?

(3) What will be the response of the chemical industry to initiatives from within marine biology? For example, if a good new microalgae source for β-carotene were found, could the market absorb the additional product or would the current chemical producers of this substance lower their prices and undercut the new technology?

(4) Will biotechnology create new products and new markets? If applied biology were merely to lower the price slightly and make slightly more available materials that are already being made in other ways, the actual contribution to the economy would not be large.

This paper will present a framework for consideration of these questions, beginning with an overview of present products and the economic constraints which apply to them. Possible areas of opportunity in commodity, specialty, and pharmaceutic chemicals will be explored. The paper will conclude by suggesting some worthwhile roles that modern biology and biotechnology could play in this area.

The possible roles of marine products and marine biotechnology can be explored by beginning with basic societal needs (Figure 1). What really could make a difference in the way the world runs? Marine biotechnology may make its greatest contribution to the production of biomass, which could be used as a feedstock for chemicals, chemical intermediates, fuels, or a combination of these. There are, of course, already very well-developed terrestrial sources of biomass. Nonetheless, there is a large commercial opportunity in this area which has so far not materialized outside of food production.

Current Production of Chemicals from Marine Sources

Table I lists the most important commercial marine products now produced, considered as organic chemicals. The most important product is clearly food, i.e., proteins. Algal

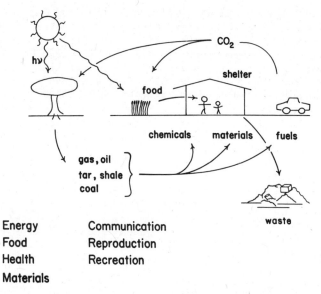

Energy   Communication
Food   Reproduction
Health   Recreation
Materials

Figure 1. Basic societal needs

Table I. Present products considered as organic chemicals

	Quantity (lb)	Price ($/lb)
Food (finfish, shellfish = protein; amino acids)	$\sim 10^{11}$	> 0.25
Algal polysaccharides (algin, carrageenan, agar)	$\sim 10^8$	3 - 7
Fish oils (fatty acids)	$\sim 10^6$	$\sim$ .135 - .15
Ethylene		$\sim$ 0.25
Glucose		$\sim$ 0.15

(Prices from Chemical Marketing Reporter)

polysaccharides (phycocolloids) are interesting and of significantly higher per weight value. Fish oils (13 to 14 cents per pound), which may presently be considered as a by-product of the fish meal industry, are less important and are one of the least valuable oil products (compare soybean oil at

18 to 20 cents per pound or peanut oil at 25 cents per pound; prices from Chemical Marketing Reporter, November 15, 1982). It is interesting to compare these figures with those for two typical, high-volume organic chemicals, ethylene and glucose. On a weight basis, prices are roughly comparable for ethylene and for fish (as food). The price for ethylene is depressed now; a more realistic price might be 28 to 29 cents per pound; likewise a real price for food might be about 30 cents per pound. An important criterion for chemicals is the price per weight of fixed carbon. On this basis ethylene is about half as expensive as fish.

By contrast, consider glucose, which is principally produced from corn. Corn-derived glucose could be considered as a material competitive with marine biomass as chemical feedstock at the current price of 15 cents per pound. Note that glucose is almost as cheap as ethylene on the basis of price per unit of carbon. Nonetheless, with the possible exception of isomerization to fructose for high-fructose syrups, glucose has not made any significant impact on the commodity chemical industry.

Marine Biomass

Marine production of biomass has qualitative characteristics which differ from those of continental biomass production. Advantages of algae biomass production include the absence of water limitation, plentiful sunlight, and perhaps most importantly, the fact that the land area used to produce algae would not be needed for other uses. Advantages of algae biomass production have been reviewed by Goldman et al. (1975), Shelef and Soeder (1980), North et al. (1981), Benemann et al. (1979), Anderson, and Richmond and Preiss (1980). It might be economically feasible, but it is not going to be politically realistic to use large areas of agricultural land to grow crops to be used as chemical feedstocks rather than food while there is starvation in the world. A serious problem for marine biomass production is nutrient availability. There are many possible solutions, including growing algae in areas of natural upwelling or using sewage or runoff; the technical details of solutions remain, nonetheless, to be worked out. Another difficulty is the nature of product harvesting. In the case of macroalgae, harvesting technology is well worked out for wild strains; however, the engineering of substrates to immobilize the seaweeds and protect them from storm and wave action is not satisfactory and this lack proscribes open ocean farming. For the microalgae, low cost harvesting techniques need to be developed because the product is very dilute. A typical phytoplankton culture reaches a concentration of a few hundred milligrams of cells per liter; by contrast, fermentations of bacteria may reach 100-200 grams per liter. Of course, sunlight is free, but we seriously question whether it might not be more practical to grow corn or trees as feedstocks for fermentations to produce chemicals than to grow algae.

Marine and continental biomass productivities, measured as
carbon fixed per unit area per day, may in some cases be quite
similar (Table 2, Ryther 1959; Lieth 1975). Under the best
circumstances sugar cane, water hyacinth, and microalgae all
produce on the order of 20 to 30 grams of dry biomass per square
meter per day, which corresponds to 5 to 10 grams of fixed
carbon per square meter per day, or an energy equivalent of
about a hundredth of a gallon of gasoline. This amount is not
enough to be wildly excited about. Under favorable circumstances marine biomass may be competitive with terrestrial
biomass, but will not be superior.

Economies of scale will be very difficult to achieve for
marine biomass production because of engineering and logistical
problems. In many processes which would use biomass for chemical production, dry product is required. Removing water is much
more straightforward for land-based agriculture than for marine
biomass production.

One final point to be addressed in marine biomass production
is the possible role of genetic engineering and other modern
biological techniques which are beginning to be applied to algae
(Sherman and Guikema 1981; Rochaix and van Dillewijn 1982).   In
the production of biomass there are typically four steps:
research (i.e., strain selection), cultivation, harvesting, and
processing. This process is shown schematically in Figure 2.
In marine biomass production one must ask what properties should
be selected during strain selection and what are the best
techniques for selecting those properties. If genetic
engineering is to be used, the question of survival in the wild

Table II. Biomass productivity

Energy Fixed as Carbon;   Biomass Productivity.

$$
\left.\begin{array}{l}\text{Sugarcane} \\ \text{Water Hyacinth} \\ \text{Macroalgae}\end{array}\right\} = \left\{\begin{array}{l} 20\text{-}30 \text{ g dry biomass}/m^2/d \\ 80\text{-}100 \text{ kcal}/m^2/d \\ \Rightarrow 0.01 \text{ gal gasoline}/m^2/d \\ \sim 2\text{-}3\% \text{ photosynthetic efficiency} \\ 5\text{-}10 \text{ g } C/m^2/d \end{array}\right.
$$

Averages

Continental	$1.8 \text{ g}/m^2/d$
Oceanic total	0.4
Upwelling Zones	1.5

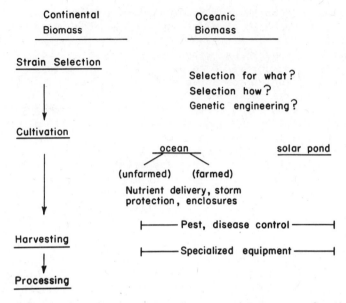

Figure 2.   Steps in developing a biomass production process

must be addressed. It is the case with bacteria that genetically engineered strains are more fragile than the wild types.
It may be necessary to control the marine environment much more
closely than presently seems practical before genetic engineering will be helpful in marine biology.

The problems discussed above are in general technically
solvable. In order to be sufficiently economically interesting
to warrant the required research and development, however,
marine biomass will require larger markets than those which are
presently available. The next section will consider the sorts
of products which might be candidates for large-scale production.

Chemicals Derived from Marine Biomass
      There are two general strategies for marine chemicals:
producing the high-value, low-volume chemicals exemplified by
algal polysaccharides, or the low-value, high-volume products
such as methane (i.e., energy). These will be considered in
turn.

      There are three commercially important algal polysaccharides
- agar and carrageenans from red algae and alginic acid from
brown seaweeds. These polysaccharides are produced in amounts -
$1-5\times10^7$ pounds per year - which are considered intermediate-
to-small quantity materials by the standards of the chemical
industry (Chapman and Chapman 1979). The world markets are on
the order of $\$10^8$ per year. This size market does not justify

a very expensive, long-term research program in biotechnology in the view of the chemical industry. It would be necessary either to expand the markets or uses or do relatively inexpensive research to increase the economics of production.

An important point about the algal polysaccharides is that their structures are complex (Figures 3, 4, and 5, Chapman and Chapman 1979; Percival and McDowell 1967). There are no competitive technologies for making these products from petroleum derived feedstocks. These products are quite secure from the standpoint of <u>direct</u> replacement by different technology, although they are, of course, vulnerable to replacement by materials with different structures and superior properties.

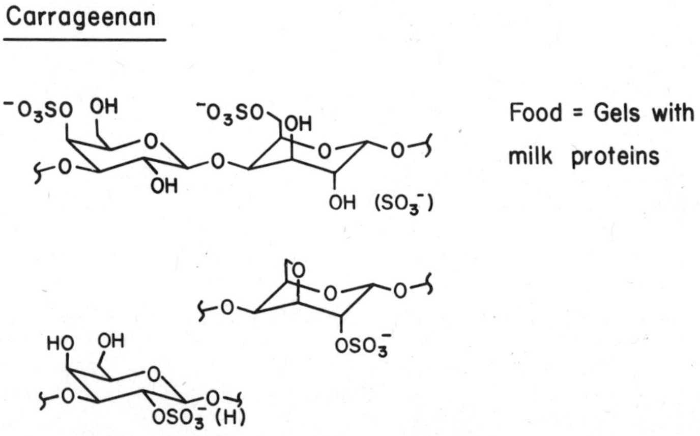

Carrageenan

Food = Gels with milk proteins

Figure 3.   Carrageenan

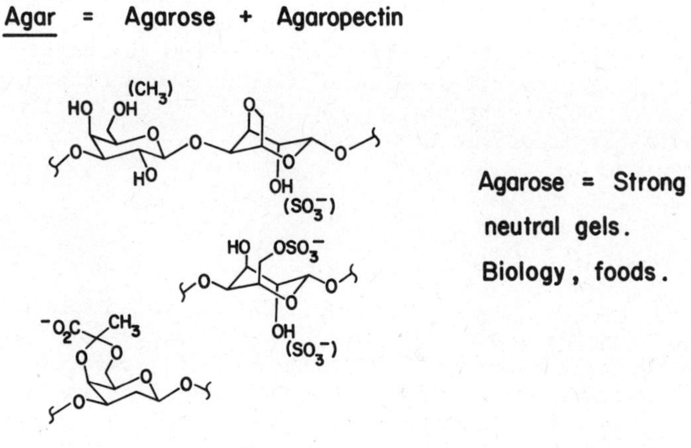

Agar = Agarose + Agaropectin

Agarose = Strong neutral gels. Biology, foods.

Figure 4.   Agar

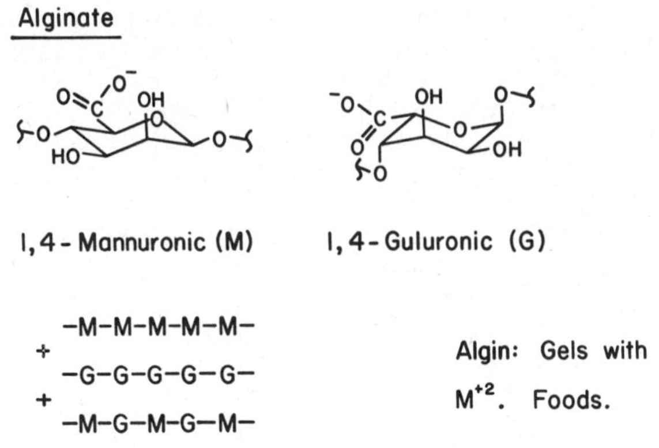

Alginate

1,4-Mannuronic (M)   1,4-Guluronic (G)

```
      -M-M-M-M-M-
   +
      -G-G-G-G-G-             Algin:  Gels  with
   +
      -M-G-M-G-M-             M⁺². Foods.
```

Figure 5. Alginate

A second favorable aspect of algal polysaccharides is that
they are used in the food industry. The food industry is tech-
nologically underdeveloped, and these products could play a
large role in a future technological expansion. The food
industry is also extremely conservative, and the fact that these
products are cleared for food use virtually guarantees their
future acceptance.

There are certain uses for the algal polysaccharides for
which there is no direct replacement. For example, carrageenan
(Figure 3) forms gels with milk proteins and so is useful in
stabilizing emulsions such as chocolate milk and ice cream (FMC
1977, 1981).

Agar (Figure 4), a related sulfated galactan, is uniquely
important to microbiology and biotechnology. Supplies of this
product are short; it is clear that world markets could absorb
considerably more than the amounts presently produced, and in
fact development of new uses would be dependent on and very
likely to follow expansion of supplies. From the point of view
of its gelling properties, the fewer sulfate and other charged
groups in the agar, the better (Guisely and Renn 1977). The
reaction below is enzyme catalyzed:

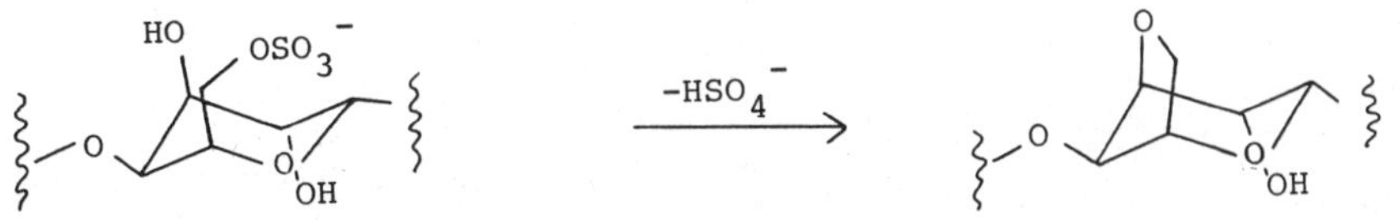

One could imagine isolating and cloning the gene for that enzyme
and developing a bioreactor to carry out this conversion for
upgrading native agar. It might be worthwhile to investigate
carefully the economics of such a scheme.

Of the present commercially important polysaccharides,
alginic acid (Figure 5) is the lowest in price. Its uses are

diverse, ranging from food to technology for immobilizing
enzymes to coatings and sizing for textile, paper, and printing
industries (Kelco 1981).

In order to justify a large research program in the area of
polysaccharides, it will be necessary to define areas of future
potential. Some specific opportunities include the following:

(1) Enhanced oil recovery is a potentially very large market
for polysaccharides if their properties are appropriate
and they can be produced cheaply enough.

(2) Certain polysaccharides, e.g., the lipopolysaccharide
emulsan from the bacterium <u>Acinetobacter calcoaceticus</u>,
could be used in petroleum transport and cleaning of
vessels and tanks.

(3) All of the algal polysaccharides mentioned here may have
expanding roles in biotechnology. Agar is used as a
culture substrate and in its highly purified form
agarose, as a separation medium. All three
phycocolloids have uses in enzyme immobilization
technology.

(4) The role of polysaccharides in food processing is likely
to increase.

(5) In any situation where the viscosity or interfacial
properties of an aqueous system must be modified or
strictly controlled, the phycocolloids can find use.
The market in such uses will depend on the relative
economics of the algal polysaccharides versus other
polysaccharides of plant or microbial origin.

Figure 6 summarizes some specific opportunities to improve
marine polysaccharides production and economics. An important
point to note is that there are many potential co-products. For
example, the brown algae which produce alginic acid also contain
a good deal of the sugar fucose in a polysaccharide called

Opportunities   (Macroalgae)

Strain Improvement :

    Classical genetics ; strain development

    Plant cell culture ⎫ To obtain what properties?

    Genetic engineering ⎭ New polymers ?

Production

    Improved farming techniques

    Enzymatic treatments

    Production engineering (supercritical fluids;
        magnetic separation)

Other

    Energy farming ?

    Coproduction of unusual sugars, lipids, enzymes.

Figure 6.   Opportunities to improve marine
polysaccharides production and economics

fucoidan. There may be a good potential market for fucose;
however, no one currently makes cheap and plentiful fucose.
Hence, market development would be required.

At the other end of the economic spectrum is methane, which
is an extremely cheap and plentiful substance. Supplies are not
unlimited, however, and schemes for methane production by
anaerobic digestion of algae have been proposed. This plan is
certainly technically feasible; various workers have shown that
many types of algae are good digestion feedstocks (Chynoweth et
al. 1981; Ryther et al. 1979; Kennan 1977; Uziel et al. 1975;
Eisenberg et al. 1980, 1981). Algae contain very little
refractory material such as lignin; water and sunlight are
available. The chief obstacle is lowering the cost of production and harvesting enough to make methane production
economical. It has been estimated that to be an economically
realistic feedstock for conversion to methane, algal biomass
must be produced and harvested at a cost of $20-30 per ton
(Dubinsky et al. 1979; Richmond and Preiss, 1980). This figure
is presently unattainable because of difficulties with nutrient
provision or harvesting. Schemes in which algae biomass
production is combined with some other valuable function, such
as waste treatment, are more likely to be successful. The
advantages and disadvantages are summarized in Figure 7.

In between the extremes of polysaccharides and methane fall
a few intermediate value products of potential interest. These
products include vegetable-like oils and fatty acids from microalgae (Shifrin and Chisholm 1980), mannitol from macroalgae
(Chapman and Chapman 1979), and glycerol from the salt-tolerant
algae Dunaliella (Ben-Amotz and Avron 1980). Some of the
Dunaliella strains contain large amounts of β-carotene; it might
be worth more than the glycerol.

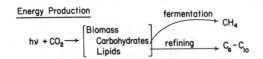

Energy Production

$$h\nu + CO_2 \rightarrow \begin{bmatrix} \text{Biomass} \\ \text{Carbohydrates} \\ \text{Lipids} \end{bmatrix} \xrightarrow{\text{fermentation}} CH_4$$
$$\xrightarrow{\text{refining}} C_6 - C_{10}$$

1) Macroalgae photosynthesize efficiently; there is
   little lignin (or equivalent) in biomass; water
   and sunlight are available.

2) Engineering of production and harvesting are
   difficult and expensive; Nutrient limited.

3) Combine with waste/sewage treatment ?
   Combine with other power-generating
         schemes ? (thermal gradient extraction ?
         Nuclear waste heat use ?)

Figure 7. Advantages and disadvantages of
marine-biomass-based energy production

Marine Pharmaceuticals

The premise in discussing pharmaceuticals is that there is a large number of new compounds in the sea, and these can be screened for useful pharmacological activity. The premise is valid, but at present there are vanishingly few human or animal drugs based on or derived from compounds first identified in marine organisms (see Fenical 1982 for a review of marine natural products chemistry and Stein and Bordon 1982 for an introduction and bibliography on algae in medicine). Many of the bioactive molecules isolated from marine sources have been toxins. This predominance of toxic substances is an artifact of the methods of bioassay (that is, will a mouse or guinea pig injected with the material die?).

Consider some interesting molecules and their possible application. Figure 8 shows the structures of palytoxin, tetrodotoxin, and brevetoxin B. These compounds are the products of very complicated biosyntheses. Genetic engineering can handle cloning and expressing the gene for one enzyme; genetically engineering an organism which could express the many enzymes which would be required to synthesize these toxins is out of the question for the next 10 to 20 years. One must, of course, ask what purpose would be served by synthesizing these molecules; the market for paralytic agents is very small!

Figure 9 shows another class of compounds of much greater commercial interest, the carotenoids. It is becoming clear that many of these compounds have interesting biological activities. The fact that these occur as photosynthetic pigments in a large number of algae (Richmond and Preiss 1980) opens up an area in which the pharmaceutic industry has specific interest. Whether this industry has the resources to go into this area now is not clear.

Some problems common to all marine pharmaceuticals are the following:
(1) Screening unfamiliar organisms may be quite difficult.
(2) There is a supposition among organic chemists that the structures found in marine species, because there is no particular advantage associated with volatility, have a tendency to be water-soluble and non-volatile, to have high molecular weight, and to be much more difficult to work with than the compounds usually handled.
(3) There is a broad question of target. If any unique marine antibacterial agents were found, they might or might not be useful against mammalian pathogens.

The marine environment certainly is worth continued examination as a source of new compounds, but it is uncertain what the payoff will be.

Animals

The largest industry at the moment having to do with the sea is the production of finfish and shellfish. Here there are many

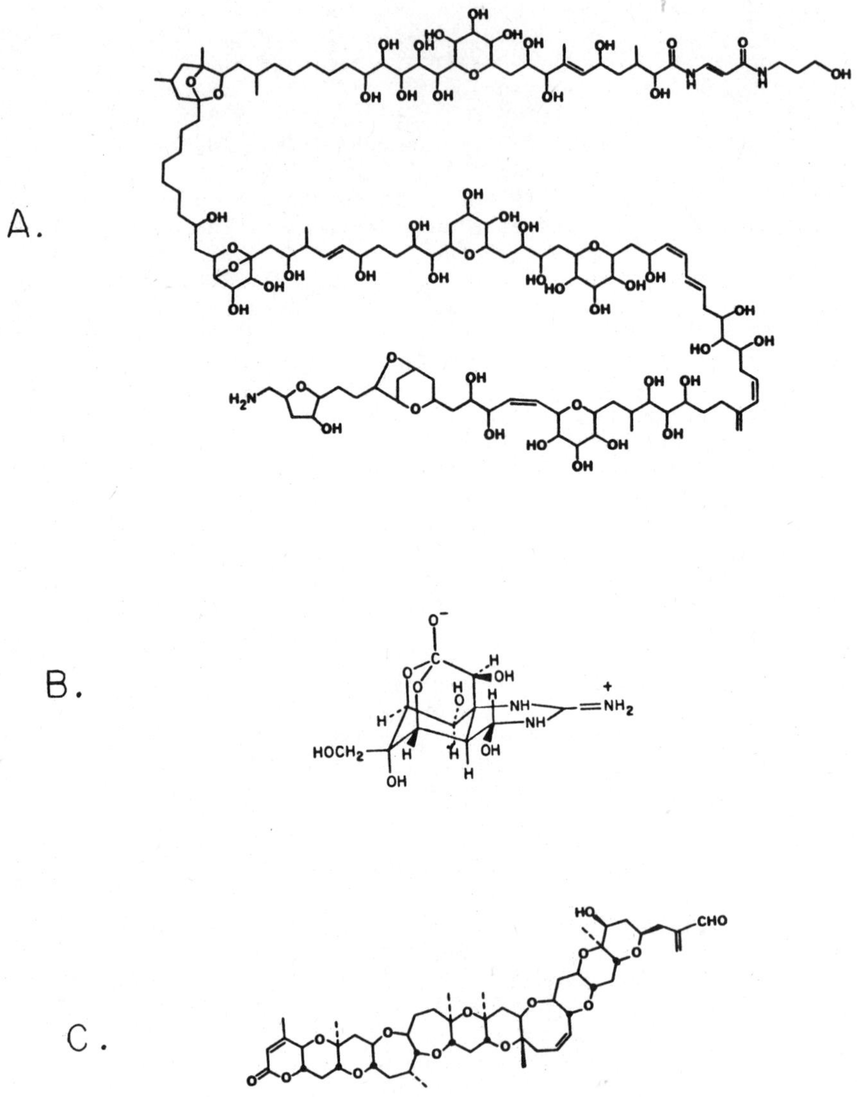

Figure 8.   (A) Palytoxin; (B) tetrodotoxin; (C)
brevetoxin B
    Source: Fenical 1982

opportunities for chemical production perhaps using biotechnology.
    The pharmaceutics markets for the aquaculture and fish –
principally trout and catfish – could expand. Chemicals for
reproduction control and synchronization of fertility in fish

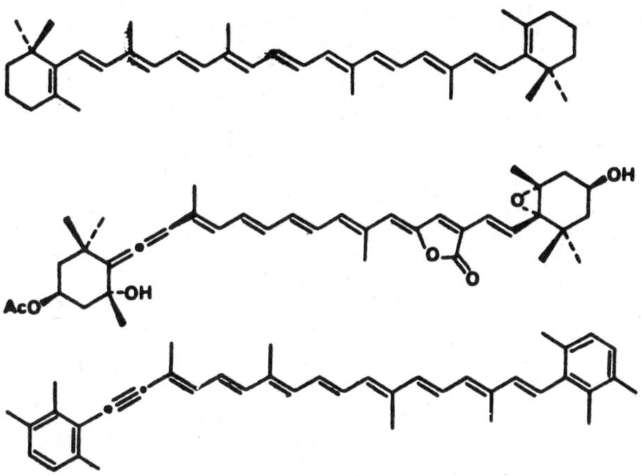

Figure 9. Carotenoids
Source: Fenical 1982

would be very useful. Communication substances, particularly to
attract fish, would be desirable. These are areas which would
require a large research effort.

There is also room for a great deal of fundamental work in
the microbiology and enzymology of spoilage and flavor
modification.

Finally, an area which clearly deserves examination is
by-product development. The viscera of fish might be used as
sources of enzymes and fine biochemicals rather than as pet
food. For example, it has been reported that carp guts contain
a small peptide factor called longevin which increases the
longevity of specific pathogen-free mice (Anon. 1982). The
structural elements, fins, scales, bones, and exoskeletons
surely contain interesting materials. An example is chitin, a
polysaccharide isolated from crab shells, which has properties
which are very distinctive and potentially useful in food,
technology, biotechnology, and medicine (Rha, this volume).

Materials Modification

Any surface put into the sea eventually serves as a
substrate for the growth of marine organisms. Any good solution
to the problem of fouling will find a huge market. Biotechnology could make a large contribution to this area, summarized in
Figure 10.

Research Opportunities

Research opportunities for organic chemicals from the sea
may be divided into relatively immediate (i.e., less than five
years), more distant, and very long term programs.

Materials Modification for Marine Use

Corrosion, Fouling

Off-shore drilling, production, mining
Thermal gradient power extraction
Floating nuclear/chemical facilities
Wave power extraction
Defense, transportation

Surface modification to control the attachment
of marine organisms. Chemical control of
surface microecology. Growth of calcareous
layers and cathodic protection.

Friction Reduction

Figure 10. Possible biotechnology contribution to marine materials modifications

Relatively immediate opportunities include improvement of existing industries:
(1) Development of useful by-products from the materials that are already collected by the fish or macroalgae industry: unusual sugars, enzymes, and other bio-chemicals. It should be possible to find high value-added biochemicals in these by-products of existing industries, in which all of the collection work has already been done.
(2) Process improvement – enzymatic polysaccharide modification of the sort discussed above and food quality maintenance and verification – present opportunities on a practical time scale.
(3) Materials development for aquaculture – especially improved fish feeds, materials for lining aquaculture tanks, and netting to which macroalgae will attach spontaneously when they are put into the water for algae farming.
The development of new products on this time scale is also possible:
(1) Sugars – It is clear that sugar chemistry is undergoing a renaissance. Everyone is considering sugars as starting materials (Fraser-Reid 1975; Fraser-Reid and Anderson 1980; Hanessian 1979; Smith and Williams 1979), and particularly as communication substances to the immune system. Materials such as fucose can have real use as chemical intermediates, if they can be provided, even at fairly high prices.
(2) Polysaccharides – Chitin is an example
(3) The carotenoids and their analogs
(4) High-value algal lipids – that is, equivalents of jojoba oil: materials which have the right molecular weight range to go into high-quality lubricating oils and into

detergent-grade alcohols. These materials are presently quite expensive, and their supply is severely limited. Land-based biotechnology is seen as a plausible strategy to provide new sources of oils. There is no reason at all why the same thing should not be practical using algae.

More distant areas (more than 5 years) for development include:

(1) Energy farming – This area will have to be coupled with something else, either alternative forms of energy production (e.g., thermal gradient power extraction) or waste disposal. The economics will not justify the development of a pure marine energy production industry, when there is so much difficulty developing the corresponding industry continentally.

(2) Pharmaceuticals – The principal interest here will be in screening biochemicals from the sea. It is important to remember that when an interesting compound is found, it is going to have to be something that can be synthesized in a laboratory, because one cannot duplicate the complex biosyntheses in microorganisms.

(3) Polysaccharides for oil production and transport.

(4) Chemicals or biotechnological processes will be required to control the problems of fouling and waste disposal.

(5) Agricultural Chemistry for Marine Systems – A fairly substantial effort in production of biologically active substances for use with marine plants and animals could be justified. The fact that this effort has not already started is a reflection of the fact that the chemical industry is only now really discovering land-based agricultural chemistry. The opportunities there and the things which are learned – particularly if one can find decent delivery systems – will justify the research in the marine environment.

Long-term opportunities are more speculative.

(1) Clearly, the sea represents a rich source of new kinds of microorganisms for screening. A tremendous untapped gene pool exists for proteins and enzymes for use in recombinant DNA. In a major sense, this potential for new proteins and enzymatic activities may prove to be / one of the most important long term contributions of marine biology to biotechnology.

(2) Another interesting area rests on the ability of many marine organisms to extract useful elements selectively from the sea, for example, diatoms concentrate silica to the extent of about 10^{12}. Almost nothing is known about these marine transport systems. Their basic microbiology may, in the long term, provide the basis for important and interesting new ways of mineral or element extractions from the oceans. For example, there was quite a seriously considered scheme for growing

Teredo worms on piles in the ocean, because the Teredo worm extracts vanadium from seawater. If one is willing to grow worms, then one should certainly be willing to work with organisms that have the appropriate characteristics to extract metals.

(3) There are whole classes of organisms that live in unusual circumstances. One of the most interesting environments is the rift region spouters, which release large amounts of H_2S. A problem of society right now is utilizing H_2S; there are not many good ways to solve this problem.

(4) Many of the marine algal systems have photosynthetic capacities which have no clear counterpart on land.

(5) Marine pheromones and communication systems would be useful for controlling the movement of marine animals during harvesting.

We see those areas as being particularly promising. It is clear that there are major opportunities, but it is also clear that they are going to require much more active exploration than the corresponding opportunities on land. It is going to behoove all of us who are interested in this area to take a very careful look at the economics of proposed targets and the downstream development after the advance biology has been done, in order to come up with the projects which justify their expenses.

References

1. Anderson, R.E. 1979. Biological Paths to Self-Reliance: A Guide to Biological Solar Energy Conversion, Van Nostrand Reinhold, New York, NY.

2. Anonymous, 1982. The elixir of life. Nature, 296:392-393.

3. Ben-Amotz, A. and M. Avron, 1980. In G. Shelef and C.J. Soeder, Eds., Algae Biomass, Elsevier/North Holland Biomedical Press, Amsterdam, pp. 603-610.

4. Benemann, J.R., J.C. Weissman, and W.J. Oswald, 1979. Algal biomass. In A.H. Rose, Ed., Microbial Biomass, Economic Microbiology, V. 4, Academic Press, New York, NY, pp. 177-206.

5. Chapman, V.S. and D.J. Chapman, 1979. The Seaweeds and Their Uses, 3rd Edition, Chapman and Hall/Methuen, New York, NY.

6. Chynoweth, D.P., S. Ghosh, and D.L. Klass. Anaerobic digestion of kelp. In S.S. Sofer and O. Zaborsky, Eds., Biomass Conversion Processes for Energy and Fuels, Plenum Press, New York, NY, pp. 315-338.

7. Dubinsky, Z., S. Aaronson, and T. Berner, 1980. Some
    economic considerations in the culture of microalgae. In G.
    Shelef and C. Soeder, Eds., Algae Biomass, Elsevier/North
    Holland Biomedical Press, Amsterdam, pp. 819-832.

8. Eisenberg, D.M., W.J. Oswald, J.R. Benemann, R.P. Goebel,
    and T.T. Tiburzi, 1980. Methane fermentation of
    microalgae. In D.A. Stafford, B.I. Wheatley, and D.E.
    Hughes, Eds., Anaerobic Digestion, Applied Science
    Publishers, Ltd., London, pp. 99-111.

9. Eisenberg, D.M., B. Koopman, J.R. Benemann, and W.J. Oswald,
    1981. Biotechnol. Bioeng. Symp., 11:429-448.

10. Fenical, W., 1982. Natural products, chemistry in the
    marine environment, Science, 215:923-928.

11. FMC, 1977. Carrageenan: Monograph Number One, FMC,
    Springfield, NJ.

12. FMC, 1981. Carrageenan: Monograph Number Two, FMC,
    Springfield, NJ.

13. Fraser-Reid, B., 1975. Some progeny of 2, 3-unsaturated
    sugars - they little resemble grandfather glucose, Acct.
    Chem. Res., 8:192-201.

14. Fraser-Reid, B., and R.C. Anderson, 1980. Prog. Chem. Org.
    Nat. Products, 39:1-61.

15. Goldman, J.C., J.H. Ryther, and L.D. Williams, 1975. Mass
    production of marine algae in outdoors culture, Nature,
    254:594-595.

16. Guisely, K.B. and D. Renn, 1977. Agarose, FMC, Springfield,
    NJ.

17. Hanessian, S., 1979. Approaches to the total synthesis of
    natural products using "chiral templates" derived from
    carbohydrates, Acct. Chem. Res., 12:159-165.

18. Kelco, 1981. Kelco Algin, Kelco, Chicago, IL.

19. Kennan, J.D., 1977. Energy, 2:265-373.

20. Lieth, H., 1975. Primary productivity in ecosystems. In
    W.H. Van Dobben and R.H. Lowe-McConnell, Eds., Unifying
    Concepts in Ecology, Dr. W. Junk B.V. Publishers, the Hague,
    pp. 67-88.

21. North, W.J., V. Gerard, and J.S. Kuwubara, 1981. Biomass

production by freshwater and marine macrophytes.   In D.
Klass, Ed., Biomass as a Non-Fossil Fuel Source, ACS Symp.
Series, 144:77-98.

22. Percival, E. and R.H. McDowell, 1967. Chemistry and
    Enzymology of Marine Algal Polysaccharides, Academic Press,
    New York, NY.

23. Richmond, A. and K. Preiss, 1980. The biotechnology of
    algaculture, Interdisciplinary Sci. Rev., 5:60-70.

24. Rochaix, J.-D. and J. van Dillewijn, 1982. Transformation
    of the green algae Chlamydomonas reinhardii with yeast DNA,
    Nature, 296:70-72.

25. Ryther, J.H., 1959. Science, 130:602-608.

26. Ryther, J.H., L. Williams, M.D. Janioak, R.W. Stenberg, and
    T.A. DeBusk, 1979. Biomass production of marine and
    freshwater plants. In 3rd Annual Biomass Energy Systems
    Conference Proceedings, SERI, Golden, CO.

27. Shelef, G. and C.J. Soeder, Eds., 1980. Algae Biomass,
    Elsevier/North Holland Biomedical Press, Amsterdam.

28. Sherman, L.A. and S.W. Guikema, 1981. Photosynthesis and
    cloning in cyanobacteria. In A. Hollander, Ed.,
    Microorganisms for Chemicals, Plenum Press, New York, NY.

Genetic Engineering Approaches in Biotechnology

Peter J. Kretschmer

Celanese Corporation
Summit, New Jersey

Abstract

Biotechnological lessons learned from well-understood
microorganisms such as <u>Escherichin</u> <u>coli</u> need to be applied to
more industrially important microorganisms. In order for
recombinant DNA technology to be applied, three problems need to
be solved for these genetically less-known, but important microorganisms: a gene ("selectable marker") is needed for identification of cells that have received exogenous DNA; DNA molecules
with a functional origin of replication must be found; and a
means must be discovered for transferring recombinant DNA molecules into the cells.

In this presentation I will look at what I consider are
three phases or levels of future research in the general area of
genetic engineering as it relates to biotechnology. I will then
look at some of the problems facing researchers working with
some less well-defined organisms, and possible new approaches to
solving the problems.

Until the 1970s the only means by which microorganisms could
be modified for improved product yield was by mutation and
strain selection. Almost invariably such yield improvement was
a hit-and-miss program, with no explanation at the molecular
level of how the mutation in question resulted in increased
yield. Thus, although considerable work and effort were put
into such strain improvement programs, they yielded very little
information that could be applied to future programs.

Possibilities with Recombinant DNA

    With the advent of recombinant DNA, two important
experimental approaches to improved product yield are now
possible: (1) specific genes can be isolated and identified,
and the concentrations of gene products per cell -- for example,
proteins that function as enzymes -- can be raised as a result
of increased gene number per cell; and (2) more important, basic
questions as to how genes are controlled, how they are switched
on and off, and how their protein products interact with and are
regulated by metabolites of the cell can be answered as a result
of the ability to isolate specific genes. The answers to these
questions, in contrast to the results obtained from mutation and
strain selection programs, can be applied to many other programs
of interest. As a result, genetic engineers can, in certain
microorganisms, increase efficiency of product yield, not just
by increasing the numbers of relevant genes per cell by cloning
the gene into a multicopy plasmid (for example, 10-fold more
genes yield 10-fold higher enzyme levels), but also by specifically altering the expression of these genes to any desired
level (that is, 10-fold more genes per cell, but also expression
of each gene 100-fold higher than normal to result in 1,000-fold
higher enzyme concentration per cell). This principle is illustrated in Figure 1. Examples of these principles are the

Cloning and Overexpression of
Metabolic Pathway Genes

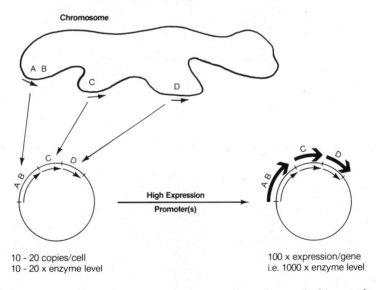

Figure 1. Cloning and overexpression of metabolic pathway
genes

recently announced overproduction of proteins such as interferon and insulin by the bacterium Escherichin coli.

In other words, genetic engineers now know how to isolate genes, study how they are expressed, and modify them appropriately. The principles for doing this at the single gene level, and mainly for one microorganism, E. coli, have now been broadly established. In a sense, that has been the first developmental phase of biotechnology.

The second phase of biotechnology will be to apply these initial lessons to the study of overexpression of and interactions between a number of genes of a particular biosynthetic or degradative pathway, as opposed to simple overproduction of any one gene. This phase will be conducted in well researched microorganisms such as E. coli and Bacillus subtilis. This second phase is more complex because (a) one is often dealing with altering the flux of critical metabolites in a cell, and (b) one must coordinate overexpression of a number of genes, and each gene must be regulated to achieve the appropriate level of overexpression.

The third and final phase of biotechnology, which will develop along with phase two, is the application of lessons learned using well understood microorganisms, such as E. coli, to the more industrially important microorganisms, such as the anaerobes, fungi, and marine organisms responsible for much of the world's important chemical activity in nature. It must be stressed, however, that the lessons learned using the well characterized microorganisms will allow much more rapid experimentation with the more industrially important microorganisms in contrast to previous programs of simple mutation and strain selection. In other words, there is a cumulative effect of the research even though it has been carried out with different microorganisms, because, in general, many of the principles of gene regulation are expected to be the same from one microorganism to another.

Researching Organisms Less-known Genetically

Most, if not all, of the organisms being discussed in this volume fall into the third phase of research. In particular, I will consider the hypothetical organism for which little if any genetic knowledge is available, for which no means of transfer of genetic material from strain to strain is available, and for which there are no immediate cloning vehicles available. How do you develop an rDNA system in a marine organism of interest, and how might a metabolic pathway be manipulated for overproduction of a desired product?

Basically, there are three problems that must be solved for establishing recombinant DNA methodology in such an organism (Figure 2). First, one needs a gene for selection, or identification, of those cells that have received the recombinant DNA of interest (i.e. a "selectable marker"); second, one needs those

Criteria for Establishment of Recombinant DNA Technology in an Organism

I. Gene for Selection
- Gene specifying antibiotic resistance
- Gene for complementation of auxotrophic marker

II. Origin of Replication
- Extrachromosomal element - plasmid
- Shuttle vectors
- Promiscuous replicons
- Chromosomal origins of replication

III. Gene Transfer System
- Classic Ca^{++} Transformation
- Microinjection
- Cell fusion
- Liposomes
- Mating

Figure 2.   Criteria for establishment of recombinant DNA technology in an organism

genes attached, or recombined, to a suitable origin of replication; and third, one needs a means of transferring this recombinant DNA molecule into the cell. These three elements -- the selectable marker, the origin of replication, and the transformation technique -- may not easily be separated when one initiates a research program on a new organism. I would now like to look at these three elements in more detail.

Broadly speaking, the gene for selection can be one of two types: either a gene specifying antibiotic resistance, or a gene capable of complementing an auxotrophic mutation in the organism. The latter requires that one first obtain an appropriate auxotrophic mutation in the organism. In fact, a selective gene such as an antibiotic resistant gene is preferable to complementation of an auxotrophic marker because the necessary auxotrophic mutation in the organism may be difficult to obtain.

The second major requirement for establishing the recombinant DNA system in a new microorganism is a DNA molecule containing a functional origin of replication (i.e., a vector). The obvious initial approach is to look for indigenous plasmids, which have been found in a wide variety of microorganisms. Once such a plasmid can be found and isolated, one can attempt to recombine it with various antibiotic resistance genes and attempt transformation into the organism. With regard to this, a variety of shuttle vectors may be useful for marine organisms. As shown in Figure 3, shuttle vectors are plasmids which are capable of replicating in two hosts, the new organism and an organism with a well established genetic knowledge base such as E. coli, B. subtilis or Saccharomyces cerevisiae. Apart from

Shuttle Vectors

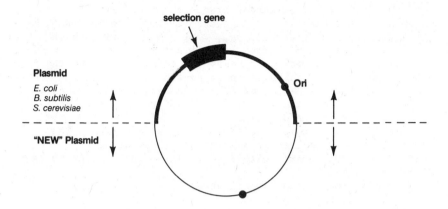

Shuttle Vectors

- Isolation of large amounts of DNA.
- Use of well established genetic system for manipulation.
- Provide possible selection gene for new plasmid.

Figure 3. Shuttle vectors

providing a ready source of large amounts of DNA from the
established organisms, shuttle vectors can instantly provide the
new plasmid with a selectable gene, depending, of course, on
whether this gene can be expressed as a functional protein
product in the new host. A recent example of this was provided
by cloning experiments in cyanobacteria, where the selectable
gene on the shuttle plasmid was expressed not only in E. coli,
but also in cyanobacteria (Kuhlemeier et al. 1981). It should
now be possible to apply the knowledge of E. coli genetics
towards the development of a genetic system in cyanobacteria.

Other sources of origins of DNA replication (i.e.,
replicons) can be promiscuous plasmids such as the plasmid RP4,
which is capable of transfer (by mating) to numerous gram-
negative microorganisms including E. coli, Salmonella, and
Pseudomonas (Barth 1979).

An important development for developing genetics in
eukaryote systems -- such as some of the marine algae --is in
the use of S. cerevisiae to isolate orgins of replications,
i.e., replicons from an organism of interest.

A selective scheme has recently been devised for isolating

DNA sequences capable of replicating autonomously in yeast, and
many such origins of replication have been isolated from such
diverse species as Neurospora, Dictyostelium, Drosophila and Zea
mays -- and of course yeast itself (Stinchcomb et al. 1980).
Very simply, one clones random fragments of DNA from the new
organism into a plasmid DNA that cannot replicate in yeast, but
which does have a selectable gene that can be expressed in
yeast. Any DNA fragment from the new organism that is capable
of supporting replication in yeast can be identified by the
appearance of yeast colonies on an appropriate selective
medium. In this manner, therefore, one can isolate a DNA
fragment from the new organism capable of replicating in yeast
and presumably in the new organism itself.
     The final requirement for developing a recombinant DNA
system in a new organism is a means of transferring the
selectable gene combined with its origin of replication to cells
of the new organisms.
     The classical procedures include:
          1. transformation
          2. transduction
          3. conjugation
It is important to recognize that new procedures have been
developed that should be applicable to the development of DNA
transfer procedures in marine organisms, and they include:
          1. protoplast fusion
          2. microinjection of DNA into the nucleus of an organism
          3. liposomes; use of DNA entrapped in inverted membrane
             vehicles which can be taken up by intact cells or
             protoplasts.
     Of course, it may not be a simple matter to satisfy these
three requirements in simple experiments. For example, even if
one has an origin of replication from the new organism and a
selective gene is tested by a classical transformation procedure, a negative result could indicate either no functional
expression of the selective gene or no successful transformation.

Establishing a Recombinant DNA System in a New Organism: An
Example in Green Algae
     Very recently, an excellent example of how one establishes a
recombinant DNA system in a new organism appeared in a report
describing transformation of the green alga Chlamydomonas
reinhardii (Rochaise and van Dillewijn 1982). The authors of
this report had the advantage of a large number of well
characterized nuclear mutants in this organism, including
arginine negative mutants. In particular, they had an
arginine-requiring mutant of C. reinhardii, for which the
analogous yeast gene of S. cerevisiae had been cloned onto a
yeast plasmid. This yeast plasmid was used to transform, with
a transformation procedure requiring the presence of diavalent
cations, the relevant C. reinhardii arginine mutant.
     This experiment had three unknowns. The first unknown was

the selective gene, because although the yeast gene coded for an analogous protein to the mutant C. reinhardii protein, it was not known if the yeast protein would be expressed in the green alga. Second, it was not known if the yeast origin of replication would function in the alga; and third, one could not be sure that the transformation conditions were correct.

Arginine positive transformants were obtained, and it was shown that the transformants contained yeast plasmid DNA. However, all the yeast plasmid DNA detected in the transformants appeared to be integrated in the host chromosome, and thus it is not clear if the yeast origin of replication is functional at all in C. reinhardii since the algae origin of replication could have been directing expression. Therefore, this system at present is unsatisfactory because although transformation is possible, one cannot recover the transformed DNA easily. Isolation of transformed DNA is required if one is to define the genes of interest and manipulate them to build up a useful knowledge base for subsequent investigation. Clearly, the next step for this study will be to obtain transformation which results from plasmids replicating independently of the host chromosome.

In summary, these types of experiments in many diverse and genetically ill-defined systems will be the focus of genetic engineering for commercially important organisms, including the marine organisms discussed in this volume. I would emphasize that recombinant DNA is only a tool in such studies -- it is still the biochemists, physiologists, and geneticists who are most important in the application of knowledge gained using recombinant DNA.

References

1. Barth, P.T., 1979. RP4 and R300B as wide host-range plasmid cloning vehicles. In K.N. Timmis and A. Puhler, Eds., Plasmids of Medical, Environmental and Commercial Importance, Elsevier/North Holland Biomedical Press, p. 399-410.

2. Goddel, D.V. et al, 1979. Expression in E. coli of chemically synthesized genes for human insulin, Proc. Nat'l. Acad. Sci. U.S.A., 76:106-110.

3. Goddel, D.V. et al., 1980. Human leukocyte interferon produced by E. coli is biologically active, Nature, 287:411-416.

4. Kuhlemeier, C.J., W.E. Borrias, C.A.M.J.J. van den Hondel, and G.A. van Arkel, 1981. Vectors for cloning in cyanobacteria: construction and characterization of two

recombinant plasmids capable of transformation to <u>E</u>. <u>coli</u> and <u>A</u>. <u>nidulans</u> R2, Mol. Gen. Genet., 184:249-254.

5. Rochaise, J.D., and J. van Dillewijn, 1982. Transformation of the green alga <u>Chlamydomonas</u> <u>reinhardii</u> with yeast DNA, Nature, 296:70-72.

6. Stinchcomb, D.T., M. Thomas, J. Kelly, E. Selker, and R.W. Davis, 1980. Eukaryotic DNA segments capable of autonomous replication in yeast, Proc. Nat'l. Acad. Sci. U.S.A., 77: 4559-4563.

Genetic Modification in Seaweeds: Applications to Commercial Utilization and Cultivation

Donald P. Cheney

Northeastern University
Boston, Massachusetts

Abstract

To get to a place you have been, you must go by a road you
have never taken.

St. John of the Cross

Man's uses of seaweeds are far more diverse and economically
important than generally realized. In the Far East, seaweeds
are intensively utilized and cultivated as a source of food. In
the Western Hemisphere, their principal commercial use is as a
source of phycocolloids agar, carrageenan, and alginate.
Although the U.S. is one of the world's largest producers and
users of phycocolloid products, it depends upon foreign resources for most of its raw material for agar and carrageenan.
While there have been numerous studies on cultivating seaweeds
in the U.S. and Canada, very little attention has been given to
producing improved strains through genetic modification outside
of the Orient. This paper will emphasize the potential role of
genetic modification in making seaweed cultivation commercially
attractive in the U.S. Through modification, or genetic engineering of existing seaweeds, it should be possible to produce
new and improved products not available in wild plants or to
reduce costs of expensive specialty products, such as agarose.
As an example of such an approach, research recently initiated
in my laboratory is briefly descibed that is aimed at producing
new, cultivatable sources of bacteriological-grade agar using

protoplast fusion-somatic hybridization techniques with cultivation, new opportunities should result for seaweed mariculture
in the U.S.

     Virtually every major agricultural crop today is a product
of some degree of genetic improvement. These improvements have
come about as a result of classical breeding techniques. Today,
a variety of new techniques, collectively referred to as genetic
engineering, are being developed and applied to plants to extend
their present limits of genetic modification and/or improvement
still further. The potential commercial applications of these
techniques to agriculture have attracted considerable attention
and investment by industry, as has been described in several
recent articles (e.g. Graft, 1982; Walsh 1981). While land
plants have been the subject of a great deal of attention,
including several symposia (e.g. Marx 1982), there has been
little, if any, attention paid to applying this new genetic
technology to their marine counterparts, the seaweeds.
     The primary focus of this paper will be to point out the
need for and potential benefits of applying genetic engineering
techniques to seaweeds. The paper's central thesis is the
belief that seaweed cultivation will only be commercially
attractive in North America once existing species can be
modified ("engineered") to produce new and improved products not
available in wild plants or produced at reduced costs. This
paper describes research currently pursued in my laboratory that
is aimed at producing new strains of agar-producing seaweeds
using protoplast fusion-somatic hybridization techniques.

Seaweed Cultivation and Utilization

     In order to understand the basic rationale behind genetic
improvement research with seaweeds, it is first necessry to
understand their worldwide commercial importance. Seaweeds are
far more economically important than generally realized. They
are used as human and animal food, in medicine and agriculture,
and as a source of raw materials for numerous industries (see
Abbott & Cheney 1982; Waaland 1981). Abbott and Cheney (1982)
have estimated that most people living in developed maritime
countries today directly or indirectly consume or come in
contact with some form of algal product daily. In many
countries of the Far East and the Pacific Islands, seaweeds
serve as an important source of food. It is estimated (Abbott
and Cheney 1982) that approximately 168 species of algae serve
as commercially important food sources worldwide, with an
estimated value of well over one billion dollars annually. Most
of this total comes from the cultivation and consumption of
Porphyra (nori), Laminaria (kombu) and Undaria (wakame) in the
countries of Japan, China, Taiwan, and Korea. The Porphyra, or
nori, industry in Japan alone is estimated to involve over
60,000 hectares in cultivation area and to be worth over $730

million annually. <u>Porphyra</u> is the most important mariculture
crop in Japan today.

In the western hemisphere, seaweeds are principally utilized
as a source of the phycocolloids agar, carrageenan and alginate
(see contribution by Renn, this volume). These three phycocolloids have a combined current world market value probably
exceeding $250 million annually. Although the U.S. is one of
the world's largest producers and consumers of phycocolloid
products, it depends upon foreign resources for most of its raw
material for agar and carrageenan.

Numerous workers have been successful in cultivating agar-
and carrageenan-producing seaweeds on a small, experimental
scale in both the United States and Canada (e.g. Cheney et al.
1981; Neish et al. 1975; DeBoer and Ryther 1978; Waaland 1979).
However, there are no commercial-scale cultivation programs to
my knowledge operating in either the U.S. or Canada today.
Commercial seaweed cultivation is, however, currently being
conducted in both enclosed systems and ocean farms in other
parts of the world (see Hansen et al. 1981; Mathieson and North
1982; Tseng 1981a).

Altogether, there are approximately 11 genera (less than 20
species) of seaweeds being cultivated to a significant commercial extent worldwide (Tseng 1981a). Most seaweed cultivation
occurs in the Orient and Indo-Pacific and includes, for example,
the large-scale farming of <u>Porphyra</u> in Japan, <u>Laminaria</u> and
<u>Undaria</u> farming in China, <u>Gracilaria</u> pond culture in Taiwan and
<u>Eucheuma</u> farming in the Philippines (Mathieson 1981; Mathieson
and North 1982; Shang 1976; Tseng 1981a). <u>Porphyra</u>, <u>Laminaria</u>
and <u>Undaria</u> are cultivated as a source of food, while <u>Eucheuma</u>
and <u>Gracilaria</u> are primarily farmed for their phycocolloids,
carrageenan and agar, respectively. Strain selection has played
an important role in the success of each of these culture
programs.

Genetic Improvement of Seaweeds

In general, the application of genetic modification/
improvement techniques to seaweeds is both recent and limited
(see Fjeld and Lovlie 1976 and Cole and Gowans 1982 for reviews
on genetic research in seaweeds in general). The most widely
used approach has been that of simple strain selection, i.e.,
the screening of wild plants for desirable traits such as fast
growth. Strain selection experiments have been conducted on
several economically important seaweeds, including <u>Chondrus</u>
(Cheney et al. 1981; Neish and Fox 1971); <u>Eucheuma</u> (Doty 1979),
<u>Gigartina</u> (Waaland 1979) <u>Porphyra</u> (Miura 1975), and <u>Laminaria</u>
and other kelps (Anon. 1976; Fang et al. 1978; Tseng 1981a).

Because of its importance as a source of carrageenan,
<u>Chondrus</u> was one of the first seaweeds in North America to be
subjected to strain selection. Largely as a result of the early
studies by A.C. Neish and co-workers at the Atlantic Regional

Laboratory (NRCC) in Halifax, Nova Scotia, methods have been
developed (and patented) for successfully growing detached
Chondrus plants in tanks using a fast growing strain referred to
as T4 (Neish and Fox 1971; Neish et al. 1975). The cultivation
potential of Chondrus has been further pursued in Canada by
Marine Colloids, Inc. (Neish 1979), a division of FMC, and in
Massachusetts by Cheney et al. (1981). In the latter study,
several fast growing strains of Chondrus were isolated from wild
populations in Canada that could be grown in the temperate
waters of Woods Hole, Massachusetts, where previous attempts to
cultivate Chondrus had been unsuccessful (Ryther et al. 1979).

     Probably the most notable success to date in the genetic
improvement of seaweeds has been that of Chinese researchers
working with the kelp Laminaria japonica, a plant not native to
Chinese waters. Through the use of a variety of techniques,
including intensive inbreeding and selection, X-ray induced
mutations and colchicine treatment, new and improved strains
have been produced that have resulted in higher yields and
extensive geographical expansion of the Laminaria culture
industry in China (Anon. 1976; Fang et al. 1978; Tseng 1981b).
The annual production of L. japonica in China has expanded from
62 fresh tons in 1952 to about 275,000 dry tons in 1979 (Tseng
1981b).

     Intra- and interspecific, as well as intergeneric,
hybridization studies have been conducted on other commercially
important seaweeds besides Laminaria (Mathieson et al. 1981).
Examples include successful crosses between species of Gigartina
(Polanshek and West 1977; West et al. 1978), Porphyra (Suto
1963), and geographically distant populations of Chondrus
crispus (Guiry 1982), as well as the unsuccessful attempt to
cross two Gracilaria species (McLachlan et al. 1977). Perhaps
the most spectacular hybridizations to date are those described
by Sanbonsuga and Neushul (1978) and Neushul (1978) involving
the kelps Macrocystis x Pelagophycus and Macrocystis x
Nereocystis. Although such studies have demonstrated the
feasibility of sexual hybridization in certain genera, they have
had limited application to seaweed cultivation to date.

     Other than the Chinese research described above, there have
been few reports of successful attempts to produce improved
strains through mutation. A notable exception is the recent
success in the laboratory of John van der Meer to produce fast
growing strains of Gracilaria tikvahiae using chemical mutation
techniques (Patwary and van der Meer, in press; van der Meer, in
press).

     It is noteworthy to keep in mind that although the degree of
genetic improvement in seaweeds has been limited to date (both
in terms of number of genera and methods), there are in fact
many places in the world where seaweeds are currently being
commercially cultivated. In view of the considerable size of
the phycocolloid industry in North America (Bixler 1979; Moss

1978), one might ask: why haven't commercial-scale cultivation programs been developed in the U.S. or Canada? The most likely explanation would appear to be that it hasn't been (or appeared) cost effective. As George Jackson concluded (Jackson 1980): "The issue is not whether seaweeds can be grown, but at what cost." Turning this argument around, it appears that for seaweed cultivation to be commercially attractive to industry in North America, it must be able to provide new and improved products not available in wild plants, or existing products at reduced cost. With this in mind, a research program was recently initiated in my laboratory aimed at developing new, fast growing strains of agar-producing seaweed that can be commercially cultivated while producing a high quality agar. Such a combination does not now appear to exist in nature and could be the basis of a new mariculture industry.

Agar Production

Because of its many important uses in bacteriological and cell culture media (see Renn, this volume), as well as a limited supply of its raw material, agar is currently the highest priced of all phycocolloids. Depending upon its quality, agar presently costs (wholesale) between $5.50–7.50 per pound for food grade and industrial grades; between $11.00–14.00 per pound for bacteriological grades (J. Moss, Agro-Mar, Inc. pers. communication 1982). Agarose, a neutrally-charged fraction of agar used in immunodiffusion and electrophoretic separations is worth between $300 per pound for some bulk quality forms and $1,600 per pound for specialized forms bought in small quantities (J. Moss, pers. communication 1982, and R. Cook, Marine Colloids, Inc., pers. communication 1982). Total worldwide production of agar was valued at nearly $48 million annually in 1976 (Moss 1978) and is probably worth over $80 million today. The bulk of the 6 to 7,000 tons of agar produced annually comes mainly from Japan and Spain with substantial amounts also produced in Argentina, Morocco, Portugal, Chile, and Korea (J. Moss, pers. communication 1982; see also Moss 1976). Although the United States is one of the largest consumers of agar (800 to 1,000 tons total usage annually; J. Moss, pers. communicaton 1982), it produces relatively little agar and, furthermore, has to import most of the raw material for what it does produce.

Agar quality varies greatly depending upon the algal species used in its preparation. Most of the raw material for bacteriological-grade, high gel-strength agar comes from species of Gelidium found along shores of Spain, Portugal, Japan, Mexico, and Morocco (Moss 1978; Yamada 1976). A few species of Gracilaria (e.g. G. verrucosa from Chile) can also produce a relatively high gel-strength agar following treatment with NaOH during extraction (Kojima and Funaki 1951). Due to over-harvesting of wild populations and the inherent slow growth of Gelidium (Wheeler et al. 1981), supplies of Gelidium in recent

years have not always been able to meet demands (Moss 1978).
Furthermore, Moss (1978) predicts that unless supplies can be
increased and stabilized, there could be a critical shortage of
high quality Gelidium agar (and therefore agarose) within just a
few years. Therefore, a stable source of bacteriological-grade
agar that can be cultivated is extremely desirable.

Agarophyte Improvement through Somatic Hybridization

Current research in my laboratory is attempting to produce
new, cultivable strains of high quality agar-producing seaweeds
using protoplast fusion-somatic hybridization techniques. A
major advantage of using somatic hybridization is that genetic
information can be transferred from one species to another
without involving (or requiring) sexual reproduction. Thus, it
is theoretically possible to hybridize individuals from different (sexually incompatible) species (or genera), as well as from
the same species (or genus). Furthermore, somatic hybridization
provides the potential to hybridize sterile individuals, as well
as species in which male and female reproductive structures are
rare and/or difficult to synchronize. Therefore, using such
techniques, it may be possible to hybridize species of
Gracilaria and Gelidium not possible through sexual hybridization.

There is a considerable body of literature available on
protoplast isolation and fusion in "higher" (i.e. seed) plants
(see Reinert and Bajaj 1977; Sharp et al 1979; Vasil 1980).
Protoplast fusion-somatic hybridization techniques have been
used successfully to produce a number of interspecific and even
intergeneric hybrids (Gleba and Hoffman 1980; Cocking 1979;
Schieder and Vasil 1980). Fusion products have been obtained
and cultured for at least a limited time from such widely
divergent species as carrot + barley, barley + soybean, barley +
tobacco, and soybean + tobacco. The fate of fusion products
appears to be at least partially affected by the relative
divergence of the parents. Intrageneric somatic hybrids have
generally demonstrated a combination of characteristics found in
both parents, regardless of their relative sexual compatibility. For example, in a hybridization of sexually incompatible
Petunia species, Power et al. (1980) found that the flower color
and morphology, leaf hair structure, chromosome number, and
isozyme patterns of the somatic hybrid were a combination of
those found in the parents. Thus, judging from experiments with
higher plant protoplasts, it appears reasonable to expect that
the somatic hybrid we hope to produce would have a combination
of parental characteristics.

To date there has been very little application of such
modern genetic improvement techniques to seaweeds. Cell and
tissue culture have been successfully conducted with the brown
algae Laminaria (Saga et al. 1978; Fries 1980) and Dictyosiphon
(Saga et al. in press), and the two red algae Chondrus (Chen and
Taylor 1978, 1979) and Gelidium (Misawa 1977). In the latter

case, Misawa (1977) reports the use of in vitro callus cultures
for the commercial production of agar, however, little mention
has been made of this work in subsequent reports.

Protoplast isolation has been accomplished in approximately
28 genera of algae (see Berliner 1981, and Adamich and
Hemmingsen 1980, for recent reviews). Most of these genera
consist of unicellular, freshwater green algae; only three
genera are not green algae and only seven genera are marine
(Berliner 1981). There is only a single report of protoplasts
being isolated from a macroscopic marine alga (Millner et al.
1979). Millner et al. (1979) successfully obtained protoplasts
from the green alga Enteromorpha, but were unable to regenerate
plants. Protoplasts have also been obtained from the sea grass,
Zostera marina (Mazzella et al. 1981). There are no reports of
protoplasts being produced from an alga as anatomically complex
as Gracilaria. Furthermore, there are no reports of protoplast
fusion in multicellular algae. However, cells of the red alga
Griffithsia have been fused in experiments utilizing empty
Nitella cells (Waaland 1978). The success of the latter study
supports the basic feasibility of protoplast fusion in red algae.

The production of a hybrid using somatic hybridization
techniques requires four basic procedures: (1) isolation and
culture of protoplasts, (2) regeneration of protoplasts into
whole plants, (3) fusion and growth of hybrid cells, and (4)
selection of hybrid cells and regeneration of hybrid cells into
hybrid plants. The basic techniques for these procedures in
seaweeds are currently being developed in our laboratory using
mutant Gracilaria tikvahiae strains (Figure 1) provided by John
van der Meer of the Atlantic Regional Laboratory.

Our basic procedures for producing protoplasts from
Gracilaria are modified after those used with microalgae and
land plants, and consist of using agarase and other cell wall
degrading enzymes (e.g. cellulysin, macerase, Onozuka R-10,
Macerozyme R-10) to hydrolyze the cell wall. Protoplasts are
isolated in a medium containing mannitol as an osmotic stabi-
lizer. (Details of the methods used will be provided in a
future report.) To date, we have been successful in producing
protoplasts from juvenile plants of Gracilaria tikvahiae, as
well as in having a limited number of protoplasts (spontaneously) fuse. Isolated protoplasts (Figure 2) have been
examined with the light microscope to make certain they lacked a
cell wall. The absence of a cell wall has been confirmed by the
lack of fluorescence following straining with Calcofluor
(Berliner et al. 1978; Berliner 1981), and by protoplasts being
osmotically fragile. Protoplasts have remained viable in culture for up to 10 days, but have so far not regenerated cell
walls. Efforts to develop a satisfactory culture medium are in
progress.

In conclusion, although protoplast research in seaweeds is
just beginning, it clearly holds great promise for increasing
seaweed utilization through strain improvement. The absence of

Figure 1. Gracilaria tikvahiae plants in culture.   x0.45

a cell wall makes protoplasts amenable to a variety of other new
genetic modification techniques in addition to protoplast
fusion, including the uptake of foreign cell organelles and DNA
encapsulated liposomes, as well as the possible use of Ti plasmids as vectors for gene transfer (Cocking 1977; Davey et al.
1980; Lurquin and Sheehy 1982). Thus, the protoplast could be a
crucial new "tool" for genetic improvement in seaweeds and could
play a critical role in their future commercial utilization and
cultivation in North America.

Acknowledgements

    The support of Biomedical Research Support Grant No. RR7143
and Northeastern University Research Scholarship Development
Fund Grants No. 7930 and 7561 is gratefully acknowledged. The
laboratory assistance of Emily Mar and plant material of
Dr. John van der Meer is gratefully acknowledged.

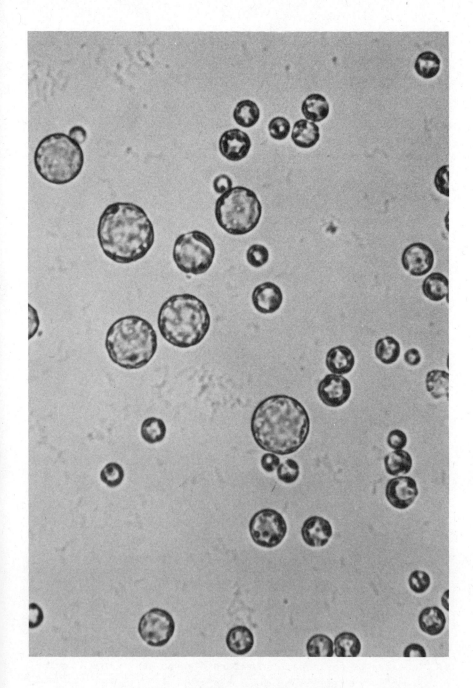

Figure 2. Freshly isolated protoplasts from G. tikvahiae
following enzymatic treatment with 1% agarase, 2% cellulysin and
1% macerase, and cultured in a seawater-mannitol medium. Small
protoplasts are from epidermal cells, large protoplasts from
cortical cells. x725.

References

1. Abbott, I.A. and D.P. Cheney, 1982. Commercial uses of
    algal products: introduction and bibliography. In J.
    Rosowski and B. Parker, Eds., Selected Papers in Phycology,
    Vol. II, Phycological Soc. Amer., Lawrence, KA, pp. 779–787.

2. Adamich, M. and B. Hemmingsen, 1980. Protoplast and
    spheroplast production. In E. Gantt, Ed., Handbook of
    Phycological Methods: Developmental and Cytological Methods,
    Cambridge Univ. Press, London, pp. 153–157.

3. Anonymous, 1976. The breeding of new varieties of haidai
    (Laminaria japonica) with high production and high iodine
    content, Sci. Sin., 19:243–252.

4. Berliner, M., N. Wood and J. Damic, 1978. Vital and
    Calcofluor staining of Cosmarium and its protoplasts,
    Protoplasma, 96: 39–46.

5. Berliner, M., 1981. Protoplasts of eukaryotic algae, Inter.
    Rev. Cytology, 73: 1–19.

6. Bixler, H., 1979. Manufacturing and marketing of
    carrageenan, In B. Santelices, Ed., Actas Primer Symposium
    Sobre Algas Marina Chilenas, Santiago, Chile, pp. 259–274.

7. Chen, L.C. and A.R. Taylor, 1978. Medullary tissue culture
    of the red alga Chondrus crispus, Can. J. Bot., 56: 883–886.

8. Chen, L.C. and A.R. Taylor, 1979. Erratum. Medullary
    tissue culture of the red alga Chondrus crispus, Can. J.
    Bot., 57:686.

9. Cheney, D., A. Mathieson, and D. Schubert, 1981. The
    application of genetic improvement techniques to seaweed
    cultivation: I. Strain selection in the carrageenophyte
    Chondrus crispus, Intl. Seaweed Symp., 10: 559–567.

10. Cocking, E., 1977. Uptake of foreign genetic material by
    plant protoplasts, Inter. Rev. Cytology, 48: 323–343.

11. Cocking. E.C., 1979. Somatic hybridization by the fusion of
    isolated protoplasts – an alternative to sex. In W. Sharp,
    P. Larson, E. Paddock and V. Raghavon, Eds., Plant cell and
    Tissue Culture Principles and Applications, Ohio State
    Univer. Press, Columbus, OH, pp. 353–369.

12. Cole, K. and C. S. Gowans, 1982. Cytology and genetics of
    algae: introduction and bibliography, In J. Rosowski and B.

Parker, Eds., Selected Papers in Phycology, Vol II,
Phycological Soc. Amer., Lawrence, KA, pp. 383-391.

13. Davey, M., E. Cocking, J. Freeman, N. Pearce and I. Tudor,
1980. Transformation of Petunia protoplasts by isolated
Agrobacterium plasmids, Plant Sci. Let., 18: 307-313.

14. DeBoer, J. and J. Ryther, 1978. Potential yields from a
waste-recycling algal mariculture system. In R. Krauss,
Ed., The Marine Plant Biomass of the Pacific Northwest
Coast, Oregon State Press, Corvalis, OR, pp. 231-249.

15. Doty, M.S., 1979. Status of marine agronomy, with special
reference to the tropics, Intl. Seaweed Symp., 9: 35-58.

16. Fang, T.C., C-H. Tai, Y-L. Ou, C-C. Tsuei, and T-C. Chen,
1978. Some genetic observations on the monoploid breeding of
Laminaria japonica, Sci. Sin., 21: 401-407.

17. Fjeld, A. and A. Lovlie, 1976. Genetics of multicellular
marine algae, In R. Lewin, Ed., The Genetics of Algae,
Botanical Monographs, Vol. 12, Univer. Calif. Press Berkley,
CA, pp. 219-235.

18. Fries, L., 1980. Axenic tissue cultures from the
sporophytes of Laminaria digitata and Laminaria hyperborea
(Phaeophyta), J. Phycol. 16: 475-477.

19. Gleba, Y.Y. and F. Hoffman, 1980. "Arabidobrassica": a
novel plant obtained by protoplast fusion, Planta, 149:
112-117.

20. Graff, G., 1982. Plant tissue culture, High Technology,
2:67-74.

21. Guiry, M., 1981. Chondrus crispus Stackhouse 'T4' is a male
clone (Rhodophyta), Phycologia, 20: 438-439.

22. Hansen, J., J.Packard and W. Doyle, 1981. Mariculture of red
seaweeds. Calif. Sea Grant College Program Publ., Report No.
T-CSGCP-002, La Jolla, CA, 42 pp.

23. Jackson, G., 1980. Marine biomass production through
seaweed aquaculture, In A. San Pietro, Ed., Biochemical and
Photosynthetic Aspects of Energy Production, Academic Press,
New York, pp. 31-58.

24. Kojima, Y. and K. Funaki, 1951. Studies on the preparation
of agar-agar from Gracilaria confervoides, Bull. Japan Soc.
Sci. Fisheries, 16: 419-422.

25. Lurquin, P. and R. Sheehy, 1982. Binding of large liposomes to plant protoplasts and delivery of encapsulated DNA. Plant Sci. Let., 25:133-146.

26. Marx, J., 1982. Plant biotechnology briefing, Science, 216: 13006-13007.

27. Mathieson, A., 1981. Seaweed cultivation: a review. In C. Sindermann, Ed., Proc. 6th U.S. - Japan Meeting on Aquaculture, U.S. Dept. Commerce Report NMFS Circ. 442, pp. 25-66.

28. Mathieson, A., and W. North, 1982. Algal aquaculture: introduction and bibliography. In J. Rosowski and B. Parker, Eds., Selected Papers in Phycology, Vol. II, Phycological Soc. Amer., Lawrence, KA, pp. 773-778.

29. Mathieson, A., T. Norton and M. Neushul, 1981. The taxonomic implications of genetic and environmentally induced variations in seaweed morphology, Bot. Rev., 47: 313-347.

30. Mazzella, L., D. Mauzerall, H. Lyman and R. Alberte, 1981. Protoplast isolation and photosynthetic characteristics of Zostera marina, Bot. Mar., 24: 285-289.

31. McLachlan, J., J. van der Meer, and N. Bird, 1977. Chromosome numbers of Gracilaria foliifera and Gracilaria sp. (Rhodophyta) and attempted hybridizations, J. Mar. Biol. Assoc. U.K., 57: 1137-1141.

32. Millner, P., M. Callow, and L. Evans, 1979. Preparation of protoplasts from the green alga Enteromorpha intestinalis, Planta, 14: 174-177.

33. Misawa, M., 1977. Production of natural substances by plant cell cultures described in Japanese patents, In W. Barz, E. Reinhard, and M. Zenk, Eds., Plant Tissue Culture and its Biotechnological Application, Springer-Verlog, Berlin, Germany, pp. 17-26.

34. Miura, A. 1979. Studies on genetic improvement of cultivated Porphyra (Laver), Proc. 7th Japan-Soviet Joint Symp. Aquaculture, pp. 161-168.

35. Moss, J., 1978. Essential considerations for establishing seaweed extraction factories, In R. Krauss, Ed., The Marine Plant Biomass of the Pacific Northwest Coast, Oregon State Press, Corvalis, OR, pp. 301-314.

36. Neish, A. and C. Fox, 1971. Greenhouse experiments on the

vegetative propagation of Chondrus crispus. Natl. Res.
Counc. Can., Atl. Reg. Lab. Tech. Rep. 12, 36 pp.

37. Neish, A., P. Shacklock, and C. Fox, 1977. The cultivation
of Chondrus crispus. Factors affecting growth under
greenhouse conditions, Can. J. Bot., 55: 2263-2271.

38. Neish, I., 1979. Principles and perspectives of the
cultivation of seaweeds in enclosed systems, In B.
Santelices, Ed., Actas Primer Symposium Sobre Algas Marina
Chilenas. Santiago,Chile, pp. 59-74.

39. Neushul, M., 1978. The domestication of the giant kelp,
Macrocystis, as a marine plant biomass producer, In R.
Krauss, Ed., The Marine Plant Biomass of the Pacific
Northwest Coast, Oregon State Press, Corvalis, OR, pp.
163-181.

40. Patwary, M. and J. van der Meer, in press. Improvement of
Gracilaria tikvahiae by genetic modification of thallus
morphology, Aquaculture.

41. Polanshek, A. and J. West, 1977. Culture and hybridization
studies on Gigartina papillata (Rhodophyta), J. Phycol., 13:
141-149.

42. Power, J.B., S.F. Berry, J.V. Chapman, and E.C. Cocking,
1980. Somatic hybridization of sexually incompatible
Petunias: Petunia parodii, Petunia parviflora, Theor. Appl.
Genet., 57: 1-4.

43. Reinert, J. and Y. Bajaj, 1977. Applied and Fundamental
Aspects of Plant Cell, Tissue, and Organ Culture,
Springer-Verlag, NY, 803 pp.

44. Ryther, J., J. DeBoer, and B. Lapointe, 1979. Cultivation
of seaweeds for hydrocolloids, waste treatment and biomass
for energy conversion, Intl. Seaweed Symp., 9: 1-16.

45. Saga, N., T. Motomura and Y. Sakai, in press. Induction of
callus from the marine brown alga Dictyosiphon
foeniculaceus, PH and Cell Physiol.

46. Saga, N., U. Takuji, and Y. Sakai, 1978. Clone Laminaria
from single isolated cell, Bull. Jap. Soc. Sci. Fish., 44:
87.

47. Sanbonsuga, Y. and M. Neushul, 1978. Hybridization of
Macrocystis (Phaeophyta) with other float-bearing kelps, J.
Phycol., 14: 214-224.

48. Schieder, O. and I. Vasil, 1980. Protoplast fusion and somatic hybridization, Intl. Rev. Cytology, Suppl., 11B:21-46.

49. Shang, Y.C., 1976. Economic aspects of Gracilaria culture in Taiwan, Aquaculture, 8: 1-7.

50. Sharp, W., P. Larson, E. Paddock, and V. Raghavon, 1979. Plant Cell and Tissue Culture Principles and Applications, Ohio State Univ. Press, Columbus, OH, 892 pp.

51. Suto, S., 1963. Intergeneric and interspecific crossings of the lavers (Porphyra), Bull. Japan Soc. Sci. Fish. 29: 739-748.

52. Tseng, C., 1981a. Commercial cultivation, In C. Lobban and M. Wynne, Eds., The Biology of Seaweeds, Botanical Monographs, Vol. 17, Univ. Calif. Press, Berkeley, CA, pp. 680-725.

53. Tseng, C., 1981b. Marine phycoculture in China, Intl. Seaweed Symp., 10:123-152.

54. van der Meer, J., in press. The domestication of seaweeds, Bioscience.

55. Vasil, I., Ed. Perspectives in Plant Cell and Tissue Culture. Intl. Rev. Cytology, Suppl. 11B, 257 pp.

56. Waaland, J., 1979. Growth and strain selection in Gigartina exasperata, Intl. Seaweed Symp., 9:241-247.

57. Waaland, J., 1981. Commercial utilization, In C. Lobban and M. Wynne, Eds., The Biology of Seaweeds, Botanical Monographs, Vol. 17, Univ. of Calif. Press, Berkely, CA, pp. 726-741.

58. Waaland, S., 1978. Parasexually produced hybrids between female and male plants of Griffithsia tenuis, a red alga, Planta, 138:65-68.

59. Walsh, J., 1981. Biotechnology boom reaches agriculture, Science, 213:1339-1341.

60. West, J., A. Polanshek and D. Shevlin, 1978. Field and culture studies on Gigartina agardhii (Rhodophyta), J. Phycol. 14:416-426.

61. Wheeler, W., M. Neushul and B. Harger, 1981. Development of a coastal marine farm and its associated problems, Intl. Seaweed Symp., 10:631-636.

62. Yamada, N., 1976. Current status and future prospects for harvesting and resource management of the agarophyte in Japan. J. Fish. Res. Board Can., 33:1024–1030.

Chitosan as a Biomaterial

ChoKyun Rha

Massachusetts Institute of Technology
Cambridge, Massachusetts

Abstract

Chitin and its deacetylated derivative chitosan are exciting polysaccharides because abundant natural resources exist and they have unique material properties. Chitin is isolated from shellfish wastes and could potentially be isolated from wastes generated by fungal fermentations; chitosan is produced by treating chitin with hot potassium hydroxide (KOH). The potential uses of chitin and chitosan are many and varied, ranging from sophisticated biotechnology applications to the treatment of waste streams. Chitosan is unusual among poly-saccharides in carrying positively charged amine groups. An in-depth analysis of the solution properties of chitosan allows prediction of its behavior in solid matrices and design of functional applications. Examples of utilization of chitosan in food products are described.

Nature of Chitosan

Chitosan is a biopolymer with unique properties which can be utilized in a variety of ways. Because of its unique and versatile properties, chitosan has great industrial potential, but this potential has yet to be exploited commercially.

Chitosan is a particularly exciting polysaccharide for three major reasons: abundant potential resources, unique material characteristics, and various functional properties.

177

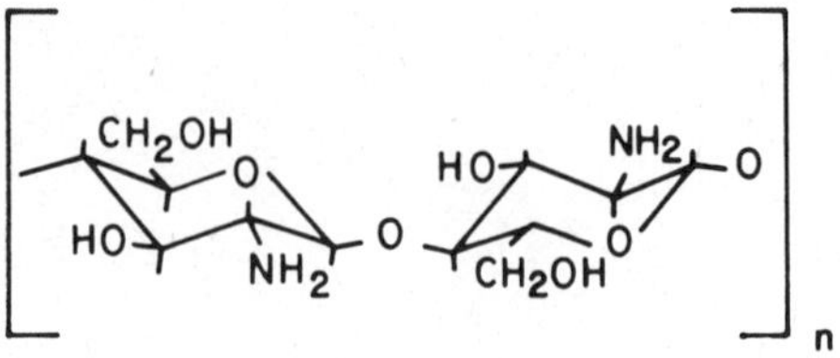

Figure 1. Chitosan

Chitosan (1-4)-2-amino-2-deoxy-β-D-glucan is a polymer with a repeating unit of disaccharides having amine groups as shown in Figure 1. It is produced currently by deacetylating chitin by boiling it in KOH or NaOH solution (Harawith et al. 1957; Braussignac 1968, Fujuta 1970).

Sources of Chitosan

Chitin, from which chitosan is derived, is the most widely distributed and one of the most abundant biopolymers in the world. The amount of chitin synthesized by marine copepods alone is estimated to be one billion tons per year (Austin et al. 1981; Berkeley 1979).

Potential commerical sources of chitin are the shells of crabs, shrimp, lobsters, krill, or clams, oysters, squid, and morays. The global annual estimate of commercial chitin production from these sources is approximately 150,000 tons per year (Allan et al. 1978; Berkeley 1979). In Alaska alone, the waste shells from the 1981 King crab processing season accounted for over 8 million pounds in less than a three month period.

For the near future, it is likely that chitin and chitosan will continue to be produced from shellfish waste. However, it is important to note another source of chitin: the fermentation industry based on fungi, which generates an estimated 800,000 tons of waste per year. About 40,000 tons of this waste is from citric acid fermentation by <u>Aspergillus niger</u> alone. This translates into approximately 10,000 tons of available chitin from this single source (Berkeley 1979; Nicolaysen 1980; Bartnicki-Garcia 1969). Therefore, there are sources of chitin other than marine, and it is feasible and perhaps economical in the near future to add to the marine resource by establishing chitin extraction or chitosan production plants in association with citric acid or antibiotics production by fermentation.

Utilization of Chitosan

Resources of chitin are abundant; there is a compelling reason to exploit them because of the wide range of functional properties of chitosan. Table I lists some common uses of chitosan. It is known to be a good adhesive and rumored to have been used to put parts of British airplane wings together during World War II. Chitosan is used to purify drinking water in

Table I. Commercial use for chitosan

Adhesive

Chelating Agent

Coagulant

Drug Carrier

Membrane

Paper and Textile Additive

Photographic Products

Surgical Adjuncts

Structural Matrix

Textile Finishers

Japan (Shinoda et al. 1980). In industrial applications, chitosan increases mechanical strength, improves dye-binding properties, reduces the consumption of sizing agents and prevents skrinkage in paper, pulp and textile products (Muzzarelli 1980; Johnson and Carlson 1978).

These varied functions of chitosan are impressive, but the application of chitosan for biomedical use and in biotechnological processes is potentially more significant and even more interesting. Also, it is worthwhile to note that such applications are more likely to be able to bear higher costs, and the current relatively high cost of chitosan would not be a deterrent for practical use (Sirica and Woodman 1970 and 1971).

In biomedical and biotechnological applications, chitosan and chitin are considered to be wound-healing agents and wound-healing accelerators (Casey 1978). Chitosan may also be used to improve the tactile properties of surgical gloves and to make suture thread, bandages and sponges (Austin 1979; Capozza 1976).

A chitosan membrane which is strong, somewhat elastic, and whose mechanical behavior resembles that of a blood vessel was fabricated in the Biomaterials Science and Engineering Laboratory at MIT. Chitosan membranes would be suitable for temporary support for skin grafting in the treatment of severe burns or in plastic surgery.

Chitosan and its derivatives are used for dialysis, deacidification of fruit, vegetables and coffee extract, and have the potential to be used in other separation processes, especially biological ones (Kesting 1979; Nakajima and Shinoda 1977; Brambilla and Horman 1980). Currently, chitosan is used in the recovery of proteins from waste streams. Chitosan has also been used for separation chromatography and to immobilize enzymes (Muzzarelli 1973, 1978).

It was proposed in the author's laboratory that micro-organisms would preferentially adhere to chitosan because it carries a net positive charge. We were, in fact, able to show that Saccharomyces cerevisiae adhere to chitosan films. We observed also that 3T3 fibroblast cells attach to chitosan films or a chitosan-carrageenan-potassium complex.

Chitosan is a material well suited for controlled release of biologically active agents and as a structural matrix. The controlled release approach to the delivery of biologically active agents such as drugs, hormones, insecticides, pesticides, and fertilizer is advantageous because the dosage can be con-trolled over a prolonged period of time; and by precise tar-geting, dosage requirements can be reduced.

We have shown that the structure of chitosan membranes can be controlled by manipulation of solution conditions (Kienzle-Sterzer et al. 1980, 1982a,b). Chitosan is biodegradable; it can be degraded by enzymes present in the normal physiological systems or in the environment. Therefore, we can provide an erosion-controlled delivery system in which release can be closely and accurately monitored. There are many more bio-medical and biotechnological applications of potential importance. For instance, antitumor activity of acetylated hexosamines has been claimed (Prudden 1976).

Properties of Chitosan in Solution

Functional performance is dictated by the fundamental material properties which make chitosan so unique and versa-tile. Chitosan has a sugar backbone, like other poly-saccharides, but it is unique in that it has amine groups which are positively charged, unlike most natural polysaccharides, which are either neutral or negatively charged (Muzzarelli 1973, 1978; Lang et al. 1982).

In order to characterize the intrinsic material properties which impart this varied functionality, one can examine the dilute solution behavior.

The flow behavior of polyelectrolyte is dictated by its overall molecular conformation and the degree of hydrogen bonding or electrostatic repulsion between neighboring chain segments. The intrinsic viscosity determined for chitosan in solution (Berkovich et al. 1980; Kienzle-Sterzer et al. 1982c; Rodriguez-Sanchez et al. 1982) indicates that chitosan adopts a conformation which ranges from random coil to a more compact "quasi-globular" shape. The intrinsic viscosity, $[\eta]$, of a linear chain polymer follows a dependence on molecular weight (MW) described in the Mark-Houwink equation (Flory 1953) where k is a constant:

$$[\eta] = k\ MW^a \tag{1}$$

For random coil polymers, the Mark-Houwink exponent, a, usually lies between 0.5 and 0.8, and the overall shape of the hydrodynamic volume is spherical. Mark-Houwink exponents greater than 0.8 reflect a highly expanded conformation which may be rod-like in shape. Values of the exponent significantly less than 0.5 reflect a more compact globular conformation. In polysaccharides, such a globular arrangement of the chain segments can be caused by local ordering due either to attractive hydrogen bonding forces or to electrostatic repulsive forces.

Table II shows the Mark-Houwink exponent, a, for chitosan and other polysaccharides in solution (Lang et al. 1982). Chitosan in dilute acid at low ionic strength has low values of a and is thus more compact than other polysaccharides due to local ordering caused by the high charge density resulting in a "quasi-globular" conformation. However, at high ionic strength and in the presence of concentrated urea, both electrostatic and hydrogen bonding forces are disrupted so that the chitosan conformation resembles that of the more typical random coil. The flexibility of the backbone chain allows chitosan to adopt the more compact conformation under low ionic strength conditions. Other charged polysaccharides such as alginate and hyaluronate have Mark-Houwink exponents approaching 1.0, reflecting a high degree of chain extension in solution.

As for all the polyelectrolyte polysaccharides, the Mark-Houwink exponents of chitosan solutions are pH- and ionic-strength-dependent, because the hydrodynamic volume is determined by the charge state of the polyion as well as by the

Table II. Mark-Houwink exponent*, a, of polysaccharides

Polysaccharide	Solvent	a
Chitosan	Trifluoracetic acid	0.296
Chitosan	Acetic acid 1%, NaCl 2.8%	0.147
Chitosan	Acetic acid 1%, LiCl 2%	0.186
Chitosan	Acetic acid 0.2M, NaCl 1M, 4M urea	0.71
Hyaluronic acid	pH 6, 0.2M NaCl	0.82
Hyaluronic acid	pH 6, 0.5M NaCl	0.78
Amylose	Water	0.5
Sodium alginate	Water	1.00
Pectin (low D.E.)	Water	0.8

*$[\eta] = K \, (MW)^a$

covalent backbone structure. Therefore, the hydrodynamic volume
of chitosan can be manipulated by altering the intramolecular
interactions between charged neighboring ions by simple changes
in solution conditions such as pH or ionic strength.

It is primarily the relative orientation of the sugar
residues around the glycosidic linkage which constrains
rearrangements in the presence of secondary forces (electrostatic forces, hydrogen bonding). The bulky nature of the sugar
rings limits rotation about the glycosidic linkage, with the
result that polysaccharide conformations in general tend to be
relatively stiff in solution. An estimation of chain stiffness
can be made using a parameter, B, (Smidsrod and Haug 1971) where

$$S = B([\eta]_I = 0.1)^{1.3} \qquad\qquad (2)$$

S is the slope of log $[\eta]$ versus log $I^{-1/2}$ where I is the
ionic strength of the solution. The constant, B, is independent
of the charge state and molecular weight of the polysaccharide
and thus provides a comparison of the backbone stiffness. The
values of B range from 0.005 for rigid rods to greater than 0.24
for polyelectrolytes with simple carbon-carbon backbones. Table
III shows values of the chain stiffness parameter B for chitosan
compared with other polysaccharides (Lang et al. 1982). Indeed,
with a B value of 0.1, chitosan has a flexible backbone compared
to most other polyelectrolyte polysaccharides such as alginate
or hyaluronic acid. This flexibility has also been evidenced by
a low degree of entanglement coupling in concentrated solution
(Kienzle-Sterzer et al. 1982d) due to decreased interpenetration
of molecular domains and by the mechanical behavior of chitosan
films (Kienzle-Sterzer et al. 1980).

The chain flexibility of chitosan makes it an especially
useful functional material because flexibility imparts an

Table III. Chain stiffness parameter of polyelectrolyte
polysaccharides

Polysaccharide	$[\eta]_I = .1$	S	B
Dextran sulfate	1.27	0.3	0.23
Carboxymethyl amylose			0.2
Chitosan	8.4	1.5	0.1
Carboxymethyl cellulose (DS = .73)	8.7	0.85	0.045
Alginate	5.9	0.42	0.04
Hyaluronic acid	8.1	0.9	0.065
Hyaluronic acid	13.4	2.2	0.07

enhanced sensitivity to charge effects in solution. The rheological properties of chitosan are extremely sensitive to variations in pH and ionic strength.

Chitosan solutions are shear thinning (Kienzle-Sterzer et al. 1982d), with the viscosity at any given concentration, pH, and ionic strength decreasing by an order of magnitude as shear rate is increased from 10 to 10^3 sec^{-1}. In the lower Newtonian region, where viscosity is independent of shear rate, the viscosity most strongly reflects the molecular conformation and the degree of chain segment interaction. However, once the low Newtonian viscosity has been established, the entire shear rate profile can be generated through the use of appropriate scaling techniques (Heldors et al. 1956; Lang et al. 1982).

At dilute solution, the intrinsic viscosity, [η], controls the lower Newtonian viscosity. Intrinsic viscosity values for chitosan are listed in Table IV along with those of several other polysaccharides (Lang et al. 1982). Increase in molecular weight, increase in the degree of ionization of the amine groups on the chitosin backbone, or decrease in the ionic strength of the solution all increase the hydrodynamic volume [η]. Table IV shows that the intrinsic viscosities of the stiff alginate and hyaluronate molecules are slightly greater than that of chitosan at comparable molecular weights. However, chitosan occupies a larger hydrodynamic volume than amylose and its derivatives, which have tightly ordered helical regions (Glicksman 1969).

The sensitivity of chitosan to ionic strength (Rodriguez-Sanchez et al. 1982) is greater than that of the other polysaccharides as evidenced by the high S value in Table III.

Matrix Characteristics
The results from the solution studies described above served as the basis for interpreting the structure/property relationships in chitosan filaments or chitosan matrices (Kienzle-Sterzer et al. 1980, 1982a,b,c,d). The chitosan matrix is simplified and presented in the schematic diagram in Figure 2. The chitosan matrix has junction zones which are mainly entanglement coupled, topologically entwined but not chemically cross-linked. At low ionic strength, the chitosan chain is extended due to electrostatic repulsion between the charged amine groups. The electrostatic repulsion orients the chains and extends the spacing between the entanglements. As ionic strength is increased, the mutual electrostatic repulsion of the amine groups is reduced, allowing the chains to become more flexible and come closer together. Further increasing the ionic strength enhances the inter- and intra-chain hydrogen bonding and causes partial collapse of the chain, which increases the number or size of junction zones provided by hydrogen bonding. Therefore, the increase in ionic strength leads to increases in the junction zone, the stiffness of the matrix and swelling, and it is likely to increase the average pore size or the fraction

Table IV. Intrinsic viscosity of chitosan and other polysaccharides.

Polysaccharide	M.W.	Solvent	$[\eta] \dfrac{dl}{gm}$	Reference
Chitosan	1.3×10^5	pH 2.5, 0.1M NaCl	8.5	Rodriguez-Sanchez et al. 1982
Chitosan	1.3×10^5	pH 2.5, 0.2M NaCl	7.7	Rodriguez-Sanchez et al. 1982
Chitosan	1.7×10^5	1% acetic acid, 2.8% NaCl	6.4	Berkovich et al. 1980
Chitosan	1.7×10^4	1% acetic acid, 2.8% NaCl	4.7	Berkovich et al. 1980
Chitosan	1.7×10^5	Trifluoroacetic acid	7.0	Berkovich et al. 1980
Chitosan	1.7×10^4	Trifluoroacetic acid	3.6	Berkovich et al. 1980
Hyaluronic Acid	5×10^5	pH 4.5, 0.1M NaCl	13.4	Lang, E.R. (1982)
Hyaluronic Acid	5×10^5	ph 4.5, 1.0M NaCl	8.5	Lang, E.R. (1982)
Amylose	1×10^6	Water	1.59	Brant, D.A. (1981)
Carboxymethyl amylose	1×10^6	0.4M NaCl	1.74	Kurata, M. and W.H. Stockmayer 1963
Alginate	1.5×10^5	NaCl	11.9	Kurata, M. and W.H. Stockmayer 1963

IONIC STRENGTH $\longrightarrow$

Figure 2.

of large pores in the matrix (Kienzle-Sterzer et al. 1982a).
This phenomenon has been observed from the dependence of residual stress on ionic strength. Increasing ionic strength produces a decrease in the residual stress, which is related to an
increase in the molecular resistance to the stress (Table V).
     Mechanical tests on the chitosan film also support the
proposed structure of chitosan films. For example, an addition
of sodium chloride increases the elastic modulus of the chitosan
matrix up to threefold from 250 g/mm^2 in the absence of salt
to 460 g/mm^2 and 757 g/mm^2 at 1.0N and 2.0N NaCl respectively (Table VI), (Kienzle-Sterzer et al. 1982a).

Applications of Chitosan
     The chain flexibility, high charge density, and hydrogen
bonding capacity of chitosan impart the ability to adopt a wide

Table V.   Volume fraction of chitosan and internal stress
parameter in swollen film

NaCl Concentration, N	Volume Fraction of Chitosan	Internal Stress g/mm^2
0.0	0.6316	5.6
0.2	0.6109	2.4
0.4	0.6061	0.2
1.0	0.6051	-1.2
2.0	0.6049	-2.6

Table VI.   Maxwell model elastic components for swollen chitosan films as a function of the ionic strength in the swelling solution[a]

NaCl Concentration N	E_1, g/mm^2
0.0	252
0.2	380
0.4	435
1.0	460
1.5	593
2.0	757

[a] 5% deformation

range of solution behavior. The solution behavior provides a basis for predicting and designing the functional performance of chitosan in fabricated forms. Some of the fundamental properties of chitosan discussed here can be used in designing and fabricating a chitosan matrix.

For example, we have fabricated chitosan globules which can have broad applications in many fields. Globules of different types were prepared by precipitation of chitosan droplets in NaOH. The globules showed a marked relationship between load and chitosan concentration and type of acid used in the casting solutions (Rodriguez-Sanchez and Rha 1981).

For food applications, opaque white globules prepared by NaOH precipitation can be used as synthetic rice devoid of calories. Globules with membranes which burst upon compression can be prepared by the precipitation of chitosan droplets in alginate solution. These globules simulate the best quality caviar, releasing the juice at the instant they are compressed (chewed!). The globules can be agglomerated, and using elongated globules one can prepare simulated fruit pulp and, in fact, synthetic oranges which contain no sugar. With much larger globules, synthetic grapes without seeds or skin can be produced.

Globules are one example in which chitosan structure can be utilized for a variety of different products. It is important to remember that the network structure of the globular membrane can be designed and fabricated according to requirements.

Our current interest in the use of chitosan globules is an encapsulation of microbial cells to be used for the production of enzymes and amino acids. The possibility of controlling the electrostatic interactions between the chitosan membrane and enzymes by manipulation of external conditions is quite exciting.

Chitosan is a biopolymer with potential to be the most important water-soluble polymer because of its abundant potential resources, unique material characteristics, and varied functional properties.

Acknowledgments
Carlos Kienzle-Sterzer, Dolores Rodriguez-Sanchez and Elizabeth Robinson Lang, graduate students in the Biomaterials Science and Engineering Laboratory, Department of Nutrition and Food Science at MIT worked on this project with undergraduate students Diane Karalekas, Robin Hoe, George Bakis and Peter Giannousis. This work was partially supported by MIT Sea Grant No. 88133, "Synthesis of Chitosan Structure Matrix for Food" and MIT Sea Grant No. 92391, "Chitosan Matrix for Biotechnology Processes" from NOAA, Department of Commerce, CONICIT of Venezuela, and ITP of Spain.

References

1. Allan, G.C., J.R. Fox, N. Kong, 1978. A critical evaluation of potential sources of chitin and chitosan, Proc. First Int. Conf. Chitin-Chitosan, p. 64.

2. Austin, P.R., 1979. U.S. Patent 4,165,433.

3. Austin, P.R., C.J. Brine, J.E. Castle, J.P. Zikakis, 1981. Chitin: new facets of research, Science, 212:749.

4. Bartnicki-Garcia, S., 1969. Cell wall chemistry, morphogenesis and toxonomy of fungi, Annu. Rev. Microbiol., 22:87.

5. Berkeley, R.C.W., 1979. Chitin, Chitosan and their degradative enzymes in microbial polysaccharides and polysaccharades. In R.C.W. Berkeley, C.W. Gooday and D.C. Elwood, Eds., the Society for General Microbiology, Academic Press, NY.

6. Berkovich, L.A., M.P. Timofeyeva, M.P. Tsyurupa, and V.A. Davankov, 1980. Hydrodynamic and conformational parameters of chitosan, Pol. Sci. USSR, 22(8):2009-2018.

7. Brambilla, A., J. Horman, 1980. Ger. Offen. 3,014,408.

8. Brant, D.A., 1981. Solution properties of polysacchardies, ACS Symposium Series, ACS, Washington, DC.

9. Broussignac, P., 1968. Un haut polymere naturel peu connu dans l'industrie: le chitosane, Chim. Ind. Genie. Chim., 99:1241.

10. Capozza, R.C., 1976. US Patent 3,989,535.

11. Casey, D.J., 1978. US Patent 4,068,757.

12. Flory, P.J., 1953. Principles of polymer chemistry, Cornell Univ. Press, Ithaca, NY.

13. Fujita, T., 1970. Japanese Patent No. 7,013,500.

14. Glicksman, M., 1969. Gum technology in the food industry, Academic Press, NY.

15. Harawith, A., S. Poseman, H.J. Blumenthal, 1957. The preparation of glucosamine oligosaccharides. I. Separation, J. Amer. Chem. Soc., 79:5046.

16. Helders, F.E., J.D. Ferry, H. Markowitz, and L.J. Zapas., 1956. Dynamic mechancial properties of concentrated solution of Na deoxyribonucleate, J. Phys. Chem., 60:1575.

17. Johnson, M.A., J.A. Carlson, 1978. Attempts to induce embrogenesis in conifer suspension cultures: biochemical aspects, Biotech. Bioengin., 20:1063.

18. Kesting, R.E., 1979. Ger. Offen. 2,849,863.

19. Kienzle-Sterzer, C.A., D. Rodriguez-Sanchez, and C.K. Rha., 1980. In G. Astarita, G. Marrucci, and L. Nicolais, Eds., Rheology Applications, Vol. 3, Plenum Press, NY, pp. 621-627.

20. Kienzle-Sterzer, C.A., D. Rodriguez-Sanchez, D. Karalekas, and C.K. Rha, 1982a. Stress relaxation of polyelectrolyte network as affected by ionic strength, Macromolecules, 15(12):631-634.

21. Kienzle-Sterzer, C.A., D. Rodriguez-Sanchez, and C.K. Rha, 1982b. Mechanical properties of chitosan films: effect of solvent acid, Makromol. Chem., 183:1353-1359.

22. Kienzle-Sterzer, C.A., D. Rodriguez-Sanchez, and C.K. Rha, 1982c. Dilute solution behavior of a cationic polyelectrolyte, J. Appl. Polym. Sci., 27:4467.

23. Kienzle-Sterzer, C.A., D. Rodriguez-Sanchez, and C.K. Rha, 1982d. Intrinsic viscosity of chitosan solutions as affected by ionic strength, Proc. Second Int. Conf. on Chitin and Chitosan, p. 26.

24. Kurata, M. and W.H. Stockmayer, 1963. Intrinsic viscosity and unperturbed dimensions of long-chain molecules, Fortschr. Hochpolym.-Forsch, Bd., 3:196-312.

25. Lang. E.R., C.A. Kienzle-Sterzer, D. Rodriguez-Sanchez, and C.K. Rha, 1982. Rheological behavior of a typical random coil polyelectrolyte: chitosan, Proc. Second Int. Conf. on Chitin and Chitosan, p. 34.

26. Muzzarelli, R.A.A., 1973. Natural Chelating Polymers, Pergamon Press, NY.

27. Nakajima, A. and L. Shinoda, 1977. Permeation properties of glycol chitosan-mucopolysaccharide complex membranes, J. Appl. Pol. Sci., 21:1249.

28. Nicolaysen, F., 1980. Chitin - a natural polymer with great industrial interest, Nature, 6:273.

29. Prudden, J.F., 1976. U.S. Patent 4,006,224.

30. Rodriguez-Sanchez, D. and C.K. Rha, 1981. Chitosan globules, J. Food Technol. (U.K.) 16:469.

31. Rodriguez-Sanchez, D., C.A. Kienzle-Sterzer, and C.K. Rha, 1982. Intrinsic viscosity of chitosan solutions as affected by ionic strength, Proc. Second Int. Conf. on Chitin/Chitosan, p. 30.

32. Shinoda, K., G. Takemura, and H. Sakimoto, 1980. Japan kokai Tokyo Koho, 80:388.

33. Sirica, A.E. and R.J. Woodman, 1971. Selective aggregation of L1210 leukemia cells by the polycation chitosan, J. Nat. Cancer Inst., 47:377.

34. Sirica, A.E. and R.J. Woodman, 1970. Aggregation of L1210 cells in vitro by chitosan and its derivatives, Federation Proc., 29:681.

35. Smidsrod, O. and H. Huag, 1971. Estimation of the relative stiffness of the molecular chain in polyelectrolytes from measurements of viscosity at different ionic strength, Biopolymers, 10:1213.

Marine Algae and Their Role in Biotechnology

Donald W. Renn

FMC Corporation
Rockland, Maine

Abstract

Because of their unique properties, products from marine
algae are playing increasingly important roles in various areas
of emerging biotechnology. Four of these products, κ-carra-
geenan, algin, agar, and agarose -- all gel-forming polysaccharides -- have found specific niches. Carrageenan and algin
can be used as cages for immobilizing cells to be used in bio-
conversions. Agar and agarose are media of choice for selection and replication of microorganisms, including plant and
animal cells. New, low gelling temperature agaroses increase
cell viability. In addition, agarose is an essential electrophoresis medium for gene mapping and separation and selection of
appropriate DNA fragments for gene splicing. Columns of beaded
agarose and its derivatives are used in biotechnology for
affinity and molecular size separations. Sources, chemical
composition, properties, and applications of these important
marine algal products are discussed.

Because of their unique properties, products from marine
algae are playing increasingly important roles in various areas
of emerging biotechnology. Four of these products -- carrageenan, algin, agar, and agarose -- are gelling polysaccharides
that have found specific niches. Before discussing these, however, I will summarize the major classes of marine algae and the
hydrocolloids extracted from them.

SEAWEEDS (ALGAE)

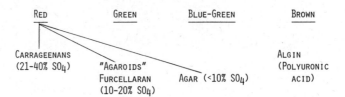

Figure 1. Seaweeds (Algae)

The four major classes of seaweeds are the <u>Rhodophyta</u> or red algae, <u>Phaeophyta</u> or brown algae, <u>Chlorophyta</u> or green algae, and <u>Cyanophyta</u> or blue-green algae (Figure 1). Only the red and brown algae are currently sources of commercial products of significant value. Three types of carrageenan, designated <u>kappa-</u>, <u>lambda-</u>, and <u>iota-</u>, and agar, from which agarose is derived by purification, are obtained from red algae, but not from the same species. Algin is obtained from a number of species of brown algae.

Carrageenan

All carrageenans are polygalactose molecules, or galactans (Figure 2). The backbone of the idealized carrageenan is an alternating $\alpha 1,3$-D galactose $\beta 1,4$-3,6 anhydro-D galactose

CARRAGEENAN STRUCTURES

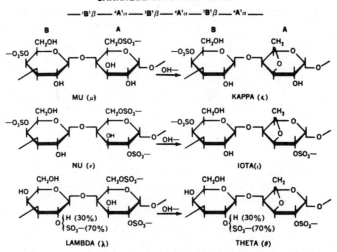

Figure 2. Carrageenan structures

subunit called 'carrabiose'. Degree of substitution and position of ester sulfate molecules determine specific types and
properties. Only the kappa-, or rigid gel forming, carrageenan,
which is sulfated in the 4 position of the galactose moiety, has
found use in biotechnology. Aqueous solutions of kappa-
carrageenan form strong, transparent, thermo-reversible gels in
the presence of potassium salts. Living or killed but enzymat-
ically active cells of yeast or bacteria, for example, can be
encapsulated or immobilized in a beaded cage of kappa-
carrageenan by introducing the sodium salt into a potassium
ion-containing solution. The beads can be used directly or
insolubilized by subsequent treatment with glutaraldehyde or
polycations and then used for bio-conversions. Dr. Ichiro
Chibata and his associates at Tanabe Seiyaku Co., Ltd. in Osaka,
Japan, have pioneered this technique and developed several
commercial processes based on it, including conversion of
glucose to ethanol and production of L-aspartic and L-malic
acids. A number of other research groups are looking into
carrageenan's potential in commercial bioconversions.

Algin

Unlike the carrageenans, whose polyanionic character is
attributable to sulfate ester groups, algin or alginic acid is
composed of variable amounts of D-mannuronic and L-guluronic
acid residues, the composition being dependent on source and
extraction procedure.

When a solution of sodium alginate is added dropwise to a
solution containing calcium ions, a water-insoluble, calcium
alginate droplet is formed. This procedure can be used to
encapsulate cells, and, as with carrageenan, these caged cells
are being evaluated as bioconversion systems. Calcium alginate
gels are not thermo-reversible but will dissolve in the presence
of a calcium-sequestering agent such as EDTA.

In addition, as was recently reported in the news media,
pancreatic cells contained in a calcium alginate sac and
implanted subdermally in a dog continued to secrete insulin for
approximately one month. Damon Corporation's new cell entrap-
ment system is reported to be algin reacted with polylysine to
decrease water solubility.

Agar

Since 1882, agar has been the medium of choice for propa-
gating and identifying bacteria, fungi, and other microorganisms, including plant and animal cells. With the advent of
gene splicing and cell fusion techniques, agar has become indi-
spensable for the selection and cloning processes.

Agar, like the carrageenans, has a polygalactose or galactan
backbone. However, unlike carrageenan, which has alternating
D-galactose / D-3,6 anhydro galactose molecules, agar is composed of "agarobiose" units with alternating D-galactose and
L-3,6 anhydro galactose (Figure 3). Ester sulfate, carboxyl,

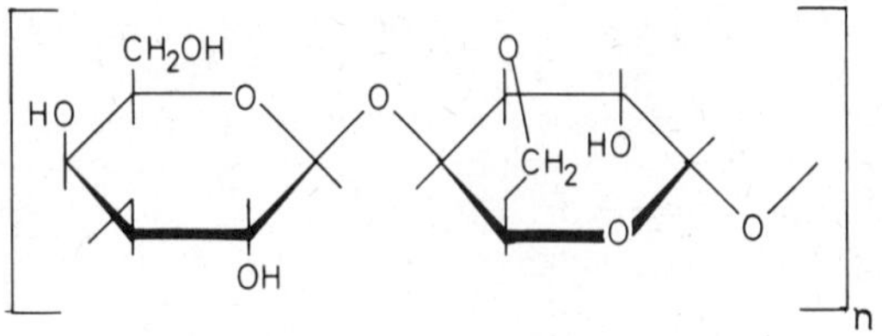

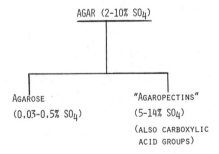

Figure 3. Agarobiose subunit

and pyruvate residues are found attached in varying positions
and quantities. Oversimplified, agar is often discussed as a
mixture of neutral or unsubstituted "agarose" and ill-defined
anionic "agarpectins." Actually, agar consists of a whole
spectrum of substituted galactans, and the properties of the
agar and agarose derived from it depend on the source and the
process used for recovery (Figure 4).

Agarose
        Last, but certainly not least in biotechnological importance
of the seaweed polysaccharides to be discussed, is agarose
(Table I). Agarose is, or I should say agaroses are, since many
types exist, the more electrically neutral, cation independent,
thermo-reversible, strong, transparent, gel-forming fractions
obtained when agar is fractionated. Although no idealized
agarose has yet been reported which does not contain anionic
substituents, some types of agarose are sufficiently devoid of
charged residues to be essentially electrically neutral and to
exhibit virtually no nonspecific protein reactivity. These can
be used as a medium on which to propagate sensitive plant,
animal, bacterial, and fungal cells. Significant yield and
percentage growth improvements over agar are being reported in
cloning procedures using agarose.
        In addition, because agarose gels water at 0.2 percent or
less, mechanically stable gels with large pores are easily
formed. These gels provide a medium for electrophoresis of

AGAR (2-10% SO$_4$)

AGAROSE   "AGAROPECTINS"
(0.03-0.5% SO$_4$)   (5-14% SO$_4$)
                                 (ALSO CARBOXYLIC
                                  ACID GROUPS)

Figure 4. Agar (2-10% SO$_4$)

Table I. Agarose: applications in biotechnology

Cell selection and propagation medium

Electrophoresis medium

      Gene mapping
      Isolating desired DNA fragments
      Product preparation

Plaquing Techniques

Column chromatography

      Gel filtration
      Affinity separations

Immobilization matrix

---

large DNA molecules. Two critical procedures in recombinant DNA techniques use agarose: (1) gene mapping and (2) separation and isolation of desired gene fragments after treatment with restriction enzymes. Because charge density is essentially equal, DNA and cleaved fragments migrate according to size in an electric field stabilized by agarose.

Hybridomas, after formation by fusion, have been found to reproduce more consistently on agarose than on agar. In this, as with many other cell culture techniques, lowered gelling temperature, hydroxyethyl agarose derivatives, such as produced by a patented process at Marine Colloids Division of FMC and given the registered names SeaPlaque® and SeaPrep® agarose, permit one to incorporate cells in a solidifying medium at temperatures well below 37°C. This increases cell viability considerably. Plaquing techniques to detect single antibody-producing cells have been simplified using low gelling temperature hydroxyethyl agarose derivatives.

Beaded agarose gels under such trade names as Sepharose® and BioGel A® are used as gel filtration media to separate products of biotechnology. Derivatized agarose beads are important for affinity column techniques. Finally, agarose immobilized cells and enzymes are important bioconverters. New biotechnology uses are continually being found for agarose and its derivatives.

Thus, as can be seen from these few examples, products from seaweeds, or marine algae, have played and are playing critical roles in many of the evolving areas of biotechnology.

Isolation and Identification of Provitamin D$_2$ and Previtamin D$_2$

Michael F. Holick

Department of Nutrition and Food Science
Massachusetts Institute of Technology
Cambridge, Massachusetts
and
Massachusetts General Hospital
Boston, Massachusetts

Abstract

It is well known that fish liver oils are a rich, natural source of vitamin D. Although it has been suggested that fish obtain this fat-soluble vitamin through the food chain, there has never been a clear demonstration that marine microorganisms have the capacity to photosynthesize vitamin D. To determine whether phytoplankton might be responsible for introducing vitamin D into the food chain two species of phytoplankton, Emiliania huxleyi and Skeletonema menzelii were grown in pure culture and then exposed to simulated solar ultraviolet radiation. Emiliania huxleyi and Skeletonema menzelii contained provitamin D$_2$ (ergosterol) and when exposed to ultraviolet radiation they both photosynthesized previtamin D$_2$ from the provitamin precursor. Therefore, phytoplankton may provide a rich untapped natural source of vitamin D and its precursors.

In the 19th century it was common folklore in northern Europe to feed children cod liver oil to prevent and cure rickets. By the turn of this century several investigators (Mellanby 1919; McCollum et al. 1921) developed a rachitic animal model and showed experimentally that cod liver oil possessed a fat-soluble factor that prevented rickets. Several years later it was demonstrated that exposure to natural sunlight or artificial ultraviolet radiation also prevented and cured this disease, and in 1929 Powers et al. (1921) demonstrated that the fat-soluble factor in cod liver oil had biological properties identical to exposure to artificial

ultraviolet radiation. In 1936 Brockmann isolated and
sructurally identified vitamin D_3 from tuna liver oil.

By the turn of the century it was clearly demonstrated that
marine phytoplankton had a high content of the fat-soluble
vitamin A. It was assumed, therefore, that these organisms also
contained the fat-soluble vitamin D and that this was the origin
of the exceptionally rich stores of vitamin D found in cod and
tuna liver oil. As early as 1927, Leigh-Clare grew the diatom
Nitzschia closterium in culture and tried to demonstrate the
presence of vitamin D without success. Several other attempts
to demonstrate vitamin D in phytoplankton were also unsuccessful
and results of doubtful significance have been reported with
zooplankton (Copping 1934).

A large number of steroids have been isolated and identified
in higher plants and algae (Goodwin, 1974). It appears that
cholesterol is widely distributed in trace amounts in higher
plants and algae. Phytosterols which are characterized by 1-C
and 2-C groups at C-24 and double bonds at C-22, C-24 (Nes and
McKean 1977; Goodwin 1974), or C-25, and by double bonds in the
steroid nucleus at positions other than C-5 are the most common
sterols found in these plants. It still remains to be determined as to whether phytoplankton are a rich source of the
D-vitamins.

Ergosterol (a $\Delta^{5,7}$-diene sterol that is the provitamin
precursor for vitamin D_2) was considered to be a major sterol
of Euglena gracilis (Stern et al. 1960; Avivi and Halevy 1967)
as well as of various Chlorophycae and red algal classes
(Beastall et al. 1971; Patterson 1971). However, some of these
observations have been disputed more recently by Brandt et al.
(1970) who demonstrated that the major free sterols in Euglena
gracilis are Δ^7-sterols rather than ergosterol, which was not
detected. Instead, these authors reported that trace amounts of
7-dehydrocholesterol and chondrillasta-Δ^5-dienol were present
in E. gracilis.

Origins of Vitamin D

My laboratory has begun a research program in collaboration
with Dr. Guillard of Bigelow Laboratories (West Boothbay Harbor,
Maine) to investigate the origin of vitamin D in the sea.
During the past five years my laboratory has elucidated the
sequential steps involved in the sun-mediated photoproduction of
vitamin D_3 in human skin (Holick 1981; Holick et al. 1980).
In order to conduct these studies (3α-^{3}H) 7-dehydrocholesterol
of high specific activity was synthesized and a new high-
resolution chromatographic system was developed which is capable
of separating 7-dehydrocholesterol (7-DHC) from its photoproducts (Figure 1). Using these tools we clearly established
for the first time that during exposure to the sun, epidermal
stores of 7-dehydrocholesterol (7-DHC) are converted to previtamin D_3 (preD_3) (Figure 2).

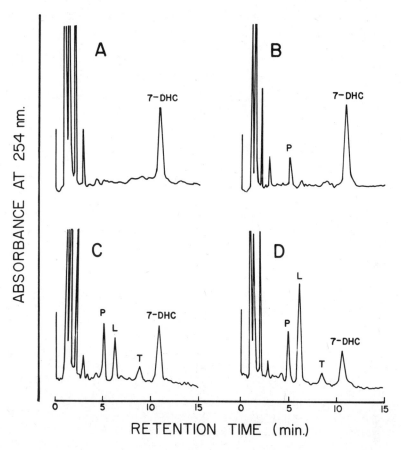

Figure 1: High-performance liquid chromatographic profile
of a lipid extract from the basal cells of surgically obtained
hypopigmented skin that was previously (A) shielded from or (B
to D) exposed to equatorial simulated solar ultraviolet
radiation that reaches Earth at sea level at noon for 10 minutes
(B), 1 hour (C), or 3 hours (D). (7-DHC = 7-dehydrocholes-
terol, preD₃ = previtamin D₃, L = Lumisterol₃, T =
Tachysterol₃.) Reproduced with permission (Holick et al. 1981).

We exposed human skin to simulated solar ultraviolet
radiation and isolated in pure form 7-DHC (also called provitamin D₃) and preD₃. These structures were identified
based on their mass spectra and ultraviolet absorption spectra.
We further demonstrated that once preD₃ was formed, this
thermally labile photoproduct underwent a temperature dependent
(1-7) sigmatropic shift of a hydrogen and an isomerization,

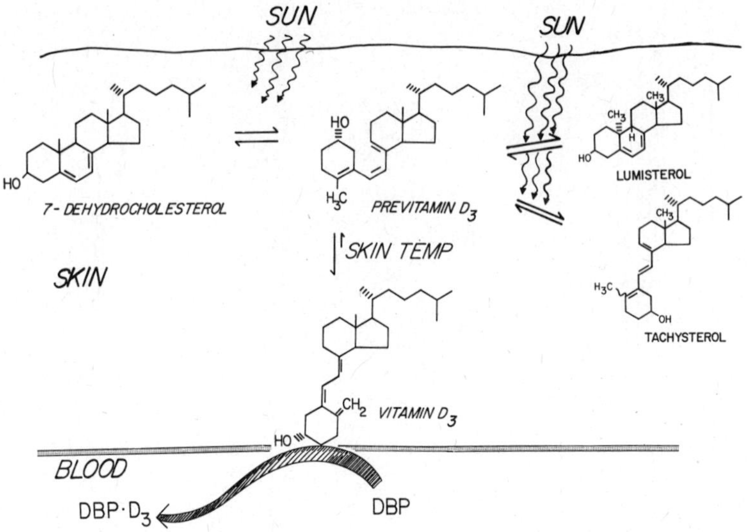

Figure 2: Schematic representation of the formation of previtamin D3 in the skin during exposure to the sun, and its thermal isomerization to vitamin D3, which is specifically translocated by the vitamin-D binding protein (DBP) into the circulation. During continual exposure to the sun, previtamin D3 also photoisomerizes to lumisterol3 and tachysterol3, which are biologically inert photoproducts; that is, they do not stimulate intestinal calcium absorption. Because the vitamin-D binding protein has no affinity for lumisterol3 but has minimal affinity for tachysterol3, the translocations of these photoisomers into the circulation is negligible, and these photoproducts are sloughed during the natural turnover of the skin. Because these photoisomers are in a quasi-stationary state, as soon as previtamin D3 stores are depleted (because of thermal isomerization to D3), exposure of lumisterol and tachysterol to ultraviolet radiation will promote the photoisomerization of these isomers to previtamin D3. Reproduced with permission (Holick et al. 1981).

within a period of three days, to vitamin D3. We also showed that there was a highly selective transport system that specifically translocated vitamin D3 from the skin into the circulation (Figure 2).

In collaboration with Dr. Robert Guillard we have used these new techniques to examine whether or not there are significant amounts of provitamin D in marine algae. We chose Emiliania huxleyi (clone BT-6) as the first species to investigate because it is a major component of oceanic plankton and because it can be easily grown in pure culture.

One hundred liters of <u>Emiliania</u> <u>huxleyi</u> grown to a cell density of about 10^6 cells/ml was produced in glass carboys as previously described (Blumer et al. 1971). The cells were harvested by centrifugation. The cells were resuspended in "Instant Ocean" (Aquarium Systems, Eastlake, Ohio), a synthetic seawater, and one half of the cells was transferred to quartz vessels and exposed to simulated solar ultraviolet radiation for various times as previously described (Holick et al. 1981).

0.025μ Ci of $(3\alpha-^3H)^7$-DHC (specific activity 4.8 Ci/mM) was added to the cells. The cells were extracted with 2:1:1 (v/v/v) methanol/chloroform/suspended-cell volume at 4°C in the dark under an inert atmosphere of argon. Twenty-four hours later one part $CHCl_3$ was added, prompting phase separation. The lower phase ($CHCl_3$) was removed and the upper aqueous phase was reextracted with 1 part $CHCl_3$. The $CHCl_3$ phases were combined, taken to dryness under N_2, redissolved in 0.1 ml of n-hexane, and applied to a straight-phase Sep-Pak preparative chromatography system (Waters Associates, Milford, MA). The column was eluted with 50 ml of n-hexane and then 50 ml of 5 percent ethylacetate in n-hexane. Five-ml fractions were collected, and 50μ 1 aliquots from each fraction were removed for tritium determinations. Tubes containing planktonic algal lipid that comigrated with $(3\alpha-^3H)7$-dehydrocholesterol were combined, and taken to dryness under N_2, redissolved in 200μ 1 of 8 percent ethylacetate in n-hexane, and chromatographed on a high-pressure liquid chromatograph (hplc) equipped with a radial compression module containing a 0.8-x 10-cm Radial Pak B column and an ultraviolet absorption detector (Model 440) at 254 nm (with a sensitivity of 0.005) connected to a printer-plotter Data Module (Waters Associates, Milford, MA). One-ml fractions were collected and 10μ 1 aliquots were taken for tritium determinations. The hplc profile (Figure 3A) demonstrated one major ultraviolet absorbing peak ($R_t10.8$ min) that comigrated with $(3\alpha-^3H)7$-dehydrocholesterol. The ultraviolet absorption spectrum of this peak as shown in Figure 3 demonstrated $\lambda_{max}295$, 282,271, which is characteristic for $\Delta^{5,7}$-diene sterols. A mass spectrum of this peak demonstrated a large number of apparent molecular ions including 384, 386, 396, and 412. It was obvious from the beginning that there would be a large number of sterols with varying sidechain structures and ring nuclear double bonds that would make up the sterol fraction from these planktonic algae (Goodwin, 1974). Because the differences in the position and number of double bonds as well as the number of additional CH_3 and small chained hydrocarbons in the sidechain do not significantly alter the polarity of the free sterol, it was not surprising to observe a single uv-absorbing peak by hplc that had an excellent ultraviolet absorption spectrum characteristic for the $\Delta^{5,7}$-diene sterols. However, a mass spectral analysis of this lipid fraction demonstrated a large number of sterols. These data are consistent with previous conflicting reports of the isolation

and identification of ergosterol (MW 396) from Euglena gracilis
by Stern (1960) and Avivi et al. (1967), and $\Delta^{5,7}$ and
Δ^7-sterols (for example, 7-DHC [MW 384] and cholestan-7-
ene-ol [MW 386]) in the same species by Brandt et al. (1970).

Although there are other highly sophisticated chromatographic techniques such as silver-nitrate impregnated silicic
acid thin-layer chromatography and gas-liquid chromatography to
separate sterols that are similar in structure, they cannot
always resolve each of the individual sterols. Furthermore,
vitamin D pyrolyzes to isopyrovitamin D and pyrovitamin D at
temperatures required for gas-liquid chromatography.

The $\Delta^{5,7}$-diene sterol, however, has a unique physical
property that can be taken advantage of for deciphering which of
the apparent molecular ions, observed for the isolated algal
lipid from hplc, is associated with the uv-absorption spectrum.
The $\Delta^{5,7}$-diene is extremely sensitive to ultraviolet radiation
(radiation spectral range 250—320 nm). When the $\Delta^{5,7}$-
chromaphore is exposed to uv-radiation it undergoes C-9 - C-10
cleavage and isomerization to form a 6,7-cis-9, 10-secosterol,
commonly referred to as previtamin D (Figure 2). As shown in
Figure 1 the previtamin D can then be easily separated from its
parent $\Delta^{5,7}$-diene sterol and thus from other contaminating
lipids that originally comigrated with the $\Delta^{5,7}$-diene sterols.

Therefore, the isolated phytoplankton $\Delta^{5,7}$-diene sterol
fraction was dissolved in diethylether and exposed to 295 ± 5 nm
radiation from a 2.5-kw xenon-mercury arc lamp coupled to a
Jobin Yvon HL-300 holographic grating monochromator for a total
dose of 5 J/cm^2 (Holick 1981; Holick et al. 1980). Immediately after exposure, the sample was taken to dryness under N_2
and applied to hplc in 8 percent ethylacetate in n-hexane. As
can be seen in Figure 3 the peak with a retention time of 10.8
min. significantly decreased (Figure 3A) and, in turn, a new
peak appeared with a retention time R_t of 5.2 min. This new
peak (R_t = 5.2 min) was isolated and shown to have an
ultraviolet-absorption spectrum with λ_{max}260 nm and λ_{min}230
nm which is characteristic of the 6,7-cis-triene for the
previtamin-D molecule (Figure 4). The mass spectrum of this
previtamin D showed a molecular ion at 396 (35 percent) and
fragments 271 (15 percent, M+—sidechain) 253 (21 percent,
271—H$_2$O), 136 (100 percent, ring A plus C-6 and C-7) and 118
(100 percent, 136—H$_2$O), which are characteristic of previtamin D$_2$. There were no higher-molecular-weight peaks or a
peak at m/e 384 (the MW of 7-DHC) to suggest that the $\Delta^{5,7}$-
diene sterol fraction from E. huxleyi was a combination of
$\Delta^{5,7}$-diene sterols. Furthermore, the peaks m/e 271 and 253
eliminated the possibility that there were any other
functionalities in the A-ring such as 4,4' substitution.
Instead, the data indicated the presence of a CH$_3$ and double
bond in the sidechain. NMR, IR, and co-chromatography studies
confirmed the structure of the isolated previtamin as previtamin

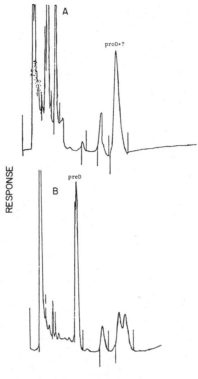

Figure 3: High-performance liquid chromatographic profiles
of lipid extracts from E. huxleyi (A) shielded from uv-radiation
or (B) exposed to uv-radiation. The chromatography was
performed on hplc with a Radial Pak B column eluted with 8
percent ethylacetate in n-hexane at 3.0 ml/min.

D$_2$. Thus the data provide incontrovertible evidence that
ergosterol was the only $\Delta^{5,7}$-diene sterol present in the
phytoplankton Emiliania huxleyi. Based upon the recovery of the
initial radioactivity that was added to the lipid extract from
E. huxleyi (the lipid solubility and hplc properties of
ergosterol and 7-DHC were identical and the radiolabelled 7-DHC
was thus an effective tracer), it was estimated that about 0.1
mg of ergosterol was present in 1 g wet weight of cells.
     Finally, to be certain that it was previtamin D$_2$ that was
isolated from the irradiated lipid extract from Emiliania
huxleyi, we converted the isolated previtamin D$_2$ to vitamin
D$_2$ by warming it at 60° for 2 hr. The isolated vitamin D$_2$
was then injected into vitamin-D-deficient rats and its biological activity was determined. The vitamin D$_2$ synthesized from

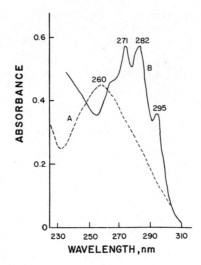

Figure 4: (A) Ultraviolet-absorption spectrum of the hplc fraction (R_T between 4 and 6 min) and (B) ultraviolet-absorption spectrum of the hplc fraction (R_T between 9 and 13 min).

the ergosterol (isolated from E. huxleyi) was capable of stimulating intestinal calcium absorption and bone calcium mobilization similar to the same dose of authentic vitamin D_2 (change in serum calcium for vitamin D_2 from E. huxleyi was 2.8 ± 0.2 mg/dl compared to 2.4 ± 0.2 mg/dl for authentic vitamin D_2.

Isolation and identification of ergosterol from Skeletonema menzelii

Using a similar approach we next investigated a semitropical phytoplankton diatom, Skeletonema menzelii, to determine whether it also has 7-dehydrocholesterol or ergosterol as its major free $\Delta^{5,7}$-diene sterol. S. menzelii was grown in pure culture as previously described. 10^9 cells were collected by centrifugation, extracted, and chromatographed as described (Blumer et al. 1971) for E. huxleyi. We isolated a $\Delta^{5,7}$-diene sterol that was converted by ultraviolet radiation to a previtamin D that had a molecular-ion and mass-fragmentation pattern that was identical to previtamin D_2. These data demonstrate that the diatom S. menzelii also makes significant quantities of provitamin D_2 (ergosterol).

The sea has been a valuable source of natural products that have important uses in biomedical science. Our observations suggest that phytoplankton may be a rich natural source of provitamin D that could be used for making vitamin D. Of equal importance is the very likely possibility that phytoplankton synthesize a variety of provitamin Ds, and, when converted to their corresponding vitamin Ds, they could have unique biologic actions in humans.

References

1. Avivi, I.O., and S. Halevy, 1967. Comp. Biochem. Physiol., 21:321.

2. Beastall, G.H., H.H. Rees, and T.W. Goodwin, 1971. Tetrahedron Letters, No. 52:4935.

3. Blumer, M., R.R.L. Guillard, and T. Chase, 1971. Marine Biol., 8:183.

4. Brandt, R.D., R.J. Pryce, C. Anding, and G. Ourisson, 1970. Z. Eur. J. Biochem., 17:344.

5. Brockmann, H., 1936. Hoppe-Seyler's Z. Physiol. Chem. 241:104-113.

6. Copping, A., 1934. Biochem. Journ., 28:1516.

7. DeSouza, N.J. and W.R. Nes, 1968. Science, 162:363.

8. Goodwin, T.W., 1974. In W.D.P. Stewart, Ed., Algal Physiology and Biochemistry, University of California Press, Berkeley, p. 266-280.

9. Holick, M.F., J.A. MacLaughlin, M.B. Clark, S.A. Holick, J.T. Potts, Jr., R.R. Anderson, I.H. Blank, J.A. Parrish, and P. Elias, 1980. Science, 210:203-205.

10. Holick, M.F., 1981. J. Invest. Derm., 76:51-58.

11. Holick, M.F., J.A. MacLaughlin, and S.H. Doppelt, 1981. Science, 211:590.

12. Kokke, W. et al. 1979. Tetrahedron Letters, 38:3601.

13. Leigh-Clare, J., 1927. Biochem. Journ., 21:368.

14. McCollum, E.V., N. Simmonds, P.G. Shipley, and E.A. Park, 1921. J. Biol. Chem., 42:507-527.

15. Mellanby, E., 1919. Lancet, 1:407-412.

16. Nes, W.R. and M.L. McKean, 1977. In Biochemistry of Steroids and Other Isopentenoids, University Park Press, p. 411-631.

17. Patterson, G.W., 1971. Lipids 6:120.

18. Powers, G.F., E.A. Park, P.G. Shipley, E.V. McCollum, and N. Simmonds, 1921. Proc. Soc. Exp. Biol. Med., 19:120-121.

19. Stern, A.I., J.A. Schiff, and H.P. Klein, 1960. J. Protozool., 7:52.

The Isolation of Labile Proteins from Marine Sources by Immobilized Monoclonal Antibodies

Gary J. Calton

Purification Engineering, Inc.
Columbia, Maryland

Abstract

Monoclonal antibodies to lethal factors in Chrysaora
quinquecirrha (sea nettle) and Physalia physalis (Portuguese
man-o-war) venom were obtained by immunization of mice with
crude venom followed by selection of appropriate hybridomas by
reactivity with selected affinity chromatography preparations
from the respective venoms. The monoclonal antibodies were
attached to Sepharose via CNBr and the resulting affinity
chromatography reagents were then used to isolate lethal
components from the venoms. In addition, novel cross-
reactivities were seen with each of these monoclonal antibodies. The monoclonal antibody to Portuguese man-o-war lethal
factor was used to isolate the crotoxin complex from the South
American rattlesnake venom. Characteristics of isolations using
immobilized monoclonal antibodies are low contamination with
extraneous protein and single step isolations of toxic materials. This work illustrates the potential for use of monoclonal
antibody isolations in isolation of proteins derived from
marine, animal, or recombinant microorganism sources.

Our research at the University of Maryland has concentrated
for some years on the venom of the sea nettle (Chrysaora
quinquecirrha). One of the most interesting facts about this
venom is that the materials contained therein are a complex
mixture of biologically active proteins: there are numerous
proteases and other degradative enzymes (Burnett and Calton
1977). The venom also contains other pharmacologically active

agents, the most important of which is the "lethal factor",
lethal to mice. The venom of the sea nettle is quite unstable
when the nematocyst (the animal's cell producing the venom) is
ruptured; we are unable to maintain the lethal character of the
venom in our laboratories for more than 24 hours at a temperature of 4°C and have had great difficulties in isolating the
active lethal factor for subsequent studies (Burnett and Calton
1977).

Isolation of the Lethal Factor

Some years ago, our lab set out to rectify this matter by
preparing monoclonal antibodies to the lethal factor of the
venom (Gaur et al. 1981). We developed an enzyme-linked
immunosorbent assay for the crude venom (Gaur, Calton, and
Burnett 1981), and immunized mice with crude venom. By taking
various chromatographic fractions from different types of
affinity chromatography columns (SP Sephadex and hexylamine
Sepharose), we were able to use the enzyme-linked immunosorbent
assays to pinpoint the antibody to the lethal factor. Using
this method, we were able to identify specific hybridomas
responsible for neutralizing the lethal factor. At this point,
it was obvious that we could use an immobilized monoclonal
antibody to isolate the lethal factor (Cobbs et al. in press).

In order to use a monoclonal antibody to isolate a component
of a mixture, the antibody must have an appropriate affinity for
the desired component. Most monoclonal antibodies actually have
a fairly low affinity for their antigen, but there are procedures that have been developed for making high-affinity antibodies when they are needed. In order to isolate a specific
component in the presence of a number of other materials, it is
necessary to attach the monoclonal antibody to a solid substrate
that may also non-specifically adsorb certain other proteins out
of the complex mixtures; thus one must choose a monoclonal antibody that will give an ideal separation. That is, affinity for
the antigen must be sufficiently high so that the antigen will
be bound significantly more tightly than other components and
yet not so high that overly harsh conditions are required to
dissociate the antibody-antigen complex.

For our purposes, we thought it would be sufficient to use a
monoclonal antibody column prepared from CNBr-activated
Sepharose-4B (Cobbs et al. in press). By selecting a monoclonal antibody with relatively low affinity for the lethal
factor, we were able to dissociate the antibody-antigen complex
with 0.75 M sodium chloride, 0.02 M phosphate, pH 7 (Figure 1).
We have worked at some length on establishing conditions necessary for dissociating this antibody-antigen complex (Cobbs et
al. in press). For example, the purity of the product dissociated with 100 mM phosphate and 10 mM sodium chloride instead
of 0.75 NaCl is not of high quality.

With the application of these recent advances in hybridoma
technology to separation science we were able to obtain 100

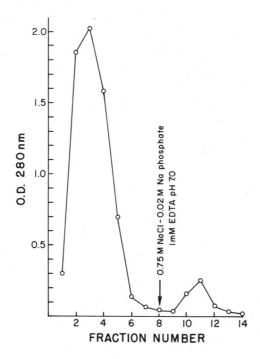

Figure 1.   Anti-sea nettle lethal factor monoclonal antibody
Sepharose column chromatography.   Sea nettle venom (SNV) protein
(15.3 mg) was applied to 3.5 ml of immunosorbent at a flow rate
of 4.0 ml/hr.   The non-complexed protein was eluted with 100 mM
NaCl-10 mM Na phosphate-1 mM EDTA pH 7.0 buffer.   The antibody-
antigen complex was dissociated with 0.75 M NaCl, 0.02 M phosphate, pH 7.   Two ml fractions were collected and tested for
lethality.

percent returns of pharmacological activity.   In previous chrom-
atographies, we had never recovered more than 30 percent of the
pharmacological activity of the starting material.   This result
is due in part to two things:
   * The mildness of the conditions for elution of the antigen
     from the antibody-antigen complex; and
   * The extremely short period of time in which affinity
     chromatographies can be conducted.
We found the crude venom (Figure 2, lane 1) could be purified (Figure 2, lane 2).   You will notice that there are two
major proteins and a number of higher molecular weight
components in the specific eluate on sodium dodecylsulfate
polyacrylamide gel electrophoresis (SDS-PAGE).   The molecular
weight of the higher band is approximately 190,000 daltons and
the lower one 100,000 daltons.   To determine whether the two
major proteins were related or one or both were impurities, we

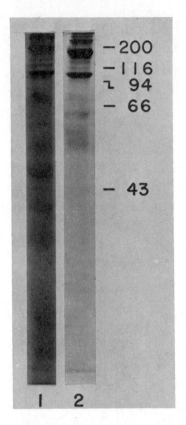

Figure 2.   SDS-polyacrylamide gel electrophoresis of sea
nettle venom (SNV) and immunosorbent column fractions.   Lane 1
contains 75 ug of crude SNV protein.   Lane 2 contains 10 ug of
purified SNLF protein from monoclonal antibody-Sepharose
column.   The molecular weights x 10-3 of the protein standards
are given to the right of lane 2.

labeled the monoclonal antibody with iodine-125, did a nitrocellulose blot, and ascertained that both of these bands reacted
with the monoclonal antibody.   This led to our belief that there
was either one site which was the same on both of these components, or that one of these two materials was derived from
degradation or dimerization of the other.   By removing these two
bands from the gel and subjecting them to limited papain
proteolysis, we were able to show that the fragments derived
from the two bands were mostly the same.   However, boiling the
bands, followed by electrophoresis, did not change the apparent
molecular weight; thus two toxin molecules are reacting with the
monoclonal antibody.

The advantage of the immunochromatography technique is that we were able to prepare pure material in large quantity. After size separation, the materials could then be used for further study in characterizing the pharmacological properties of the "lethal factors". As Table I notes we obtained 33 percent of the total lethal activity in the specific eluate. Other lethal factors are present within the venom and we are presently trying to isolate those. In addition, by having the monoclonal antibody to the lethal factor available, we were able to examine its cross reactivities with other biologically active materials.

Cross Reactivities Found

Unexpected results were obtained with the monoclonal antibody, actually in a serendipitous manner, because we were looking at the allergic response of patients to the venoms of the sea nettle. A significant cross reactivity of the monoclonal antibody to the sea nettle was found with the venom of the Portuguese man-o-war (Physalia physalis) and the venom of the sea wasp (Chironex fleckerii), a jellyfish found off the northern coast of Australia. Sea wasp venom is lethal to man and has caused death in as little as 48 seconds. A cross reactivity was also found with the venom of a wasp, the Oriental hornet (Vespa orientalis), and the venom of the South American rattlesnake, Crotalus durissus terrificus (Russo et al. in press).

In Figure 3 the SDS PAGE of the results of immunochromatography of C. d. terrificus is shown. The crotoxin complex, which is only with some difficulty dissociated, is indicated. We put Crotalus durissus terrificus venom (lane A) on our immobilized monoclonal antibody column and isolated the crotoxin

Table I. Purification of lethal factor from sea nettle fishing tentacle nematocyst venom

Stage	Total Protein ug	LD_{100} Activity ug/g	Recovery of Lethal %	Fold Increase in Activity
Venom	11,000	2.78	100	1.0
Monoclonal antibody-Sepharose 0.75 M NaCl, 0.02 M phosphate pH7 eluate	620	0.37	33	7.4

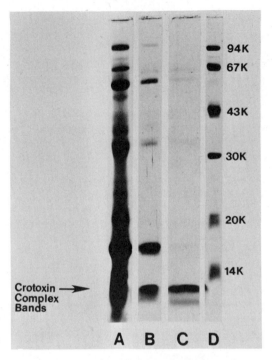

Figure 3. SDS polyacrylamide gel electrophoresis of
Crotalus durissus terrificus venom fractions from anti-
Portuguese man-o-war antibody-Sepharose column. Lane A, crude
rattlesnake venom, 100 ug protein; lane B, non-retained protein
(15 ug) eluate; lane C, 0.1 phosphate buffer eluate (10 ug
protein); lane D, molecular weight standards, phosphorylase B
(94,000), bovine serum albumin (67,000), ovalbumin (43,000),
carbonic anhydrase (30,000), soybean trypsin inhibitor (20,100)
and -lactalbumin (14,400). The two protein bands of the
crotoxin complex are identified.

complex (lane C), which consists of two bands on SDS PAGE (Russo
et al. in press). As can be seen, we have found some unexpected
relationships between various marine and other animal venoms.

We believe that these data serve as a model system to show
the potential for purification via immobilized monoclonal
antibodies. A number of companies have been founded to develop
this technology for purification. The main problem still
present in these separations is extraneous protein adsorption to
the polymer matrix on which the monoclonal antibody is immobilized. Purification Engineering, Inc. was set up specifically
to build polymer matrices which do not have the problems
associated with monoclonals immobilized on Sepharose. These
problems include poor flow-rate, compaction, non-specific

protein adsorption, and loss of the immobilized antibody due to nucleophilic reaction of the isourea bond formed by cyanogen bromide.

We think that the labile marine venoms may well serve to open up new vistas in biotechnology as models for the isolation of very complex and labile proteins from a large number of other sources.

I should like to thank my co-workers J.W. Burnett, C.S. Cobbs, A.J. Russo and P.K. Gaur for their contributions to this work.

References

1. Burnett, J.W. and G.J. Calton, 1977. The chemistry and toxicology of some venomous pelagic coelenterates, Toxicon, 15: 177.

2. Cobbs, C.S., P.K. Gaur, A.J. Russo, J.E. Warnick, G.J. Calton and J.W. Burnett, in press. Immunosorbent chromatography of sea nettle Chrysaora quinquecirrha venom and characterization of toxins.

3. Gaur, P.K., G.J. Calton and J.W. Burnett, 1981. An enzyme-linked immunosorbent assay to detect antibody against sea nettle venom, Experentia, 37: 1005.

4. Gaur, P.K., R.L. Anthony, T.S. Cody, G.J. Calton and J.W. Burnett, 1981. Production of a monoclonal antibody against the sea nettle venom mouse lethal factor, Proc. Soc. Exp. Biol. Med., 167: 374.

5. Russo, A.J., C.S. Cobbs, G.J. Calton, J.W. Burnett, in press. Detection of common antigenic sites in lethal proteins of non-related animals venoms.

Shark Cartilage Contains an Inhibitor of Tumor Neovascularization

Anne Lee and Robert Langer

Massachusetts Institute of Technology
Cambridge, Massachusetts

Abstract

Cartilage is a source of potentially useful biochemicals, including a substance proven extremely effective at inhibiting vascular proliferation in tumors. Because sharks have a large endoskeleton composed almost entirely of cartilage, they may be a unique source of the inhibitor. Efforts to test the efficacy of these substances have been constrained by the small yield of inhibitor from mammalian cartilage.

The purpose of this paper is to explore the possibility that (a) substances exist which inhibit neovascularization, (b) these substances can be used to restrict the growth of solid tumors, and, finally, (c) marine organisms, in particular sharks, are a source of such chemicals.

Tumors grow in a small three-dimensional state until they get to be about a millimeter in diameter. Without blood vessels, they would stop at this size. The problem is one of nutrition. Cells in the center begin to die because they cannot get oxygen or nutrients. Cells on the outside will continue to proliferate. At about 10^6 cells, or 1mm diameter, equilibrium is reached, and without blood vessels the tumor grows no further. However, tumors are able to produce a chemical substance known as tumor angiogenesis factor (TAF), which sends out a signal to new blood vessels and causes abnormal capillary proliferation toward the tumor. Once the new blood vessels enter the tumor, it becomes very large (Folkman and Cotran 1976).

Neovascularization and Tumor Growth

One example of the importance of neovascularization in relation to tumor growth can be seen in experiments in which a tumor was placed in the anterior chamber of the eye, which doesn't have blood vessels. When the tumor was avascular, no growth past a 1 mm diameter was observed. However, once the tumor becomes vascularized -- as ascertained by fluorescein injection-- extremely rapid tumor growth was observed. From day 5 after implantation to day 14, a 16,000-fold increase in tumor volume was noted (Gimbrone et al. 1972).

A question of interest to us was how the vascularization process could be stopped. It is known that cartilage in the embryonic stage is vascularized, whereas in a newborn, it is not vascularized (Haraldson 1962). It occurred to us that cartilage might contain certain neovascularization inhibitors. To test this possibility, cartilage from newborn rabbits and veal was used. Experiments were conducted where a tumor was placed in the cornea of a rabbit. A boiled tissue (control) or cartilage (experiment) was placed adjacent to the tumor.

In the controls, (V2 carcinoma and boiled cartilage), the blood vessels had grown by day 10 from the edge of the cornea towards the boiled cartilage. By day 20, the vessels had entered the tumor, and it had become vascularized. By 30 days, a very large, three-dimensional tumor was observed and the animal was sacrificed. In contrast, in the experimental corneas, the vessels did not grow as quickly. The vessels were sparser and formed a zone of inhibition around the cartilage (Brem and Folkman 1975).

The next step was to obtain an abstract that we could test. We had obtained approximately 100 tons of veal bones over several years, extracted and partially purified a fraction, and infused it into animals. In one experiment, we weighed tumors seven days after implantation. The controls weighed 41 times more than the tumors in the treated animals (Langer et al. 1980). Similar results were obtained in different models and with different ways of delivering the inhibitor (Langer et al. 1976 and 1980).

However, there was a major limitation. Because of low yields of inhibitor, we could test only a few animals a year with all of the tons of veal bones used.

Shark Cartilage as Inhibitor

It occurred to us that sharks, unlike mammals, have a skeleton which is entirely cartilagenous and might be an abundant source of this inhibitor. We took 2,000 grams of basking shark fin, scraped off the flesh from the cartilage, cut the cartilage into small strips and extracted it with 1 molar guanidine for 6 weeks. The extract was then dialyzed against water to remove the guanidine. This caused some reaggregation. The soluble material was then lyophilized yielding 4.79 grams. To perform a bioassay, we again used the rabbit cornea (Figure

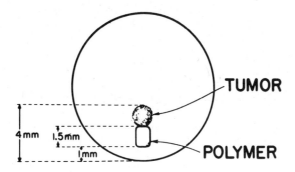

Figure 1.

1). A V2 carcinoma was placed into the cornea. Next to it was
placed a slow-release polymer which is capable of delivering a
macromolecule, such as a protein or a proteoglycan, for over 100
days (Langer and Folkman 1976). The bioassay involved measuring
the length of the longest blood vessel to the tumor.

In controls, by day 15, hundreds of blood vessels had grown
from the edge of the cornea towards the tumor. Five days later,
a very large three-dimensional tumor was observed. However,
when we tested the extract from shark cartilage the density of
blood vessels was much lower. The vessels formed a zone of
inhibition around the polymer. The length of the longest blood
vessels was 70-80 percent lower than that in the controls
(Figure 2).

We have also found other substances in shark cartilage,
including a potent cell growth factor, lysozyme, and protease
inhibitors. Thus, these studies provide initial evidence that
shark cartilage may be a source of biologically active polypeptides.

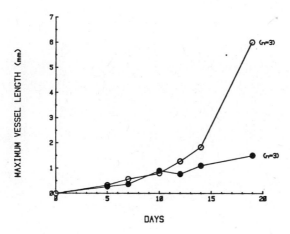

Figure 2.

REFERENCES

1. Brem, H., and J.J. Folkman, 1975. Inhibition of Tumor Angiogenesis Mediated by Cartilage, J. Exp. Med., 141:427.

2. Folkman, J. and R.S. Cotran, 1976. Relation of Vascular Proliferation to Tumor Growth, Int. Rev. Exp. Pathol, 16:207.

3. Gimbrone, M.A., Jr., S. Leapman, R.S. Cotran, and J. Folkman, 1972. Tumor Dormancy In Vivo by Prevention of Neovascularization, J. Exp. Med., 136:261.

4. Haraldson, S., 1962. The Vascular Pattern of a Growing and Fullgrown Human Epiphysis, Acta Anat., 48:156.

5. Langer, R., H. Brem, K. Falterman, M. Klein, and J. Folkman, 1976. Isolation of a Cartilage Factor which Inhibits Tumor Neovascularization, Science, 193:70.

6. Langer, R. and J. Folkman, 1976. Polymers for the Sustained Release of Proteins and Other Macromolecules, Nature, 263:797.

7. Langer, R., H. Conn, J. Vacanti, C. Haudenschild, and J. Folkman, 1980. Control of Tumor Growth in Animals by Infusion of an Angiogenesis Inhibitor, Proc. Nat. Acad. Sci., 77:4331.

III.   MARINE BIOFOULING

Microbial Ecology of Biofouling

Rita R. Colwell

University of Maryland
College Park, Maryland

Abstract

From the marine engineer's point of view, marine biofouling
is destructive to vessels and structures in seawater, but the
bacteria that cause biofouling are important components of the
natural aquatic ecosystem. Among the beneficial roles of biofouling is the enhancement and control of set by the larvae of
invertebrates in hatchery tanks. The goal of research in biofouling, therefore, should be the control, not the elimination
of biofouling. The prospects for genetic control mechanisms as
an approach to regulating biofouling are developing rapidly.
Important functions in the attachment/relationship of bacteria
with their hosts are emerging from research in several laboratories: survival of microorganisms, geographic distribution of
bacteria, and larval attraction are examples of these functions. The process of biofouling on various types of surfaces
and materials is also being studied.

During the past decade, microbial primary film formation has
been investigated by microbiologists, with the result that the
details of the process involved in biofouling have been elucidated. The role of flagella in attachment of bacteria to
surfaces has proven fascinating (Belas and Colwell 1982a, b).
Molecular genetic aspects of flagella formation and attachment
are now being investigated by Melvin Simon and his colleagues at
the Agouron Institute in LaJolla, California, (see Simon, this
volume) among others, and the prospects for genetic control

mechanisms to be used as an approach to the control of biofouling are exciting, indeed.

Aquatic bacteria are a significant component of natural
aquatic ecosystems. These bacteria are of ecological significance, not only as primary colonizers of complex substrates,
such as organic particulates in the water column and sand grains
on beaches, but they also carry out heavy metal transformations,
degradation of complex organic compounds, nutrient cycling, and
mineralization. Very frequently the heavy metal-resistant
bacteria that mediate heavy metal transformations also are
antibiotic-resistant (Allen et al. 1977). Furthermore, aquatic
bacteria are important components in the food chain (Berk and
Colwell 1981; Marshall 1976).

In this presentation, the bacterial view, or perspective, of
attachment and its function in biofouling will be presented.
From the marine engineer's viewpoint, biofouling is destructive,
a process to be blocked completely. However, a balanced,
ecological view is required. Indeed, the aquatic bacteria and
fungi are responsible for major steps in corrosion of metals and
destruction of wood and other materials in the marine environment. Yet, the attachment process and related microbial
activities are of fundamental importance to the individual
microorganisms if they are to survive, replicate, and colonize
throughout the environment. From that perspective, it is
important to gain an understanding of biofouling as one of the
many fundamental processes to be controlled, but not eliminated. The forces of nature and of evolution are such that total
elimination of the biofouling process is to eliminate an
integral component of the ecosystem. Thus, selective control
should be the goal of biofouling research and technology.

Any surface, whether a lobster held in a tank or in nature,
or a wooden piling, when exposed to natural waters, whether
seawater, brackish, estuarine, or freshwater, will immediately
be colonized. The extent of colonization will vary from a few
bacteria and/or fungal spores initially, to heavy colonization
with eucaryotic organisms over time. A single blade of turtle
grass, for example, can support a complex epibiotic community,
comprised of many billion organisms representing several taxa.

A major component of the biota of the Chesapeake Bay, or any
estuary for that matter, is the plankton, notably zooplankton.
From our studies of bacteria-copepod relationships in the
Chesapeake Bay, as shown in scanning electron micrographs of
Eurytemora, the surfaces of copepods (Figure 1), especially
around the oral region (Figure 2), are known to support significant numbers of attached bacteria. In fact, the copepod can
feed on bacteria (Berk and Colwell 1981). The relationship of
the copepod and bacteria is quite specific. That is, certain
groups of bacteria have evolved a natural association with
copepods specifically. The gut flora of copepods can be almost
entirely vibrios (Sochard et al. 1979). This association
appears to be a commensal relationship between bacteria and

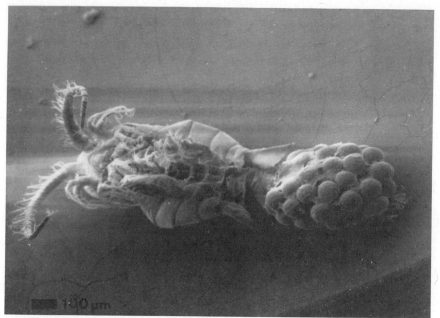

Figure 1. Scanning electron micrograph of a female copepod. Arrows denote the oral region and the egg case.

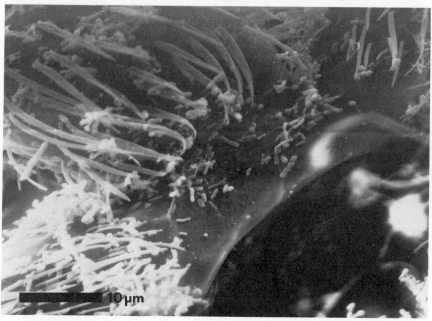

Figure 2. Scanning electron micrograph of the oral region of a live copepod colonized with _Vibrio cholerae_ serovar 01.

copepods, and results to date suggest that an osmoregulatory
interaction occurs between certain of the vibrios and their host.

An important function of attachment appears to be survival
of the microorganism. In the case of <u>Vibrio cholerae</u>, found by
us to be a naturally-occurring organism in the estuarine
environment (Colwell et al. 1981), when it is in association
with live copepods, it will survive indefinitely (Huq 1983).
Dead copepods, on the other hand, do not offer a selective
advantage for the vibrio, and survival time is short. Clearly,
there is a significant increase in survival of <u>V. cholerae</u> when
in association with living copepods. Colonization of dead
copepods by bacteria does occur, but it is quite non-specific
(Figure 3), and attachment in this case results in mineralization of the dead animal.

Interestingly, the egg case of the gravid female copepod is
extensively colonized by vibrios, with preferential attachment
to the egg sac (Figure 4). Since the eggs of copepods are dispersed in the water column, attachment provides a mechanism for
extended geographic distribution of the bacteria, yet another
function of attachment.

The life cycle of the copepod and seasonal distribution of
vibrios occurring in association with copepods appears to be
inextricably linked (Kaneko and Colwell 1973). The vibrios grow

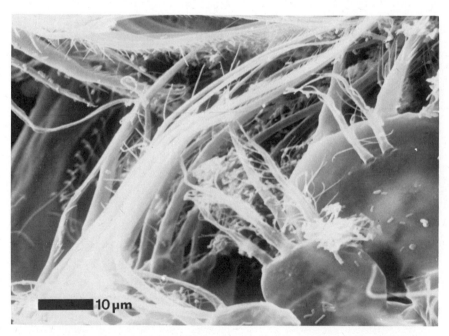

Figure 3. Scanning electron micrograph of the oral region
of a dead copepod showing scattered, non-specific attachment of
bacteria.

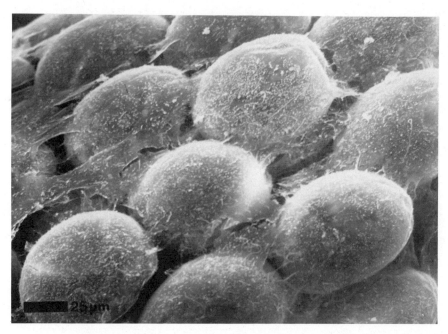

Figure 4. Scanning electron micrograph of an egg case from
a live female copepod colonized with Vibrio cholerae serovar O1.

in association with the copepod, as described above, and are
released into the water column via egg dispersal and also
subsequent to the plankton "bloom" when there is a die-off of
the copepod, i.e., its natural cycle (Kaneko and Colwell 1973).
     Another important function of the attachment/relationship of
bacteria with their hosts that is becoming more clear as a
result of research in several laboratories, including that of
Mitchell at Harvard and Weiner at the University of Maryland, is
that of larval chemotaxis or attraction. The attachment of
larvae seems to be enhanced, if not completely regulated, by
bacterial film formation. In the case of barnacles, sponges,
oysters, and related organisms in the marine environment,
settlement is enhanced and metamorphosis induced by products of
an associated bacterium.
     In the case of the larvae of the oyster, Crassostrea
virginica, of commercial importance in the Chesapeake Bay, an
interesting relationship with microorganisms has been discovered
(Weiner et al. 1983). In hatchery tanks at the University of
Delaware where larvae are raised, a bacterial film is formed on
the materials used to collect spat. A bacterium has been
isolated which is responsible for the film formed on the spat
collector surface. Without the bacterial film, far fewer spat
are collected. The bacterium produces a pigment, a melanin-like
compound, and the pigment alone can induce increased attachment

of the spat. Coating the spat collector with dead bacterial
cells yields a greatly reduced spat set, essentially the same as
a surface without biofilm formation. Clearly, here is yet
another role for bacterial film formation, i.e., "biofouling",
namely enhancement and perhaps control of set by larvae of
invertebrate animals.

The melanin pigment produced by the bacterium common to the
Delaware hatchery is also readily isolated from areas of the
Chesapeake Bay where good spat set is achieved. The pigment
precursor, L–DOPA, a neurotransmitter, is also effective in
attracting oyster larvae. Interestingly, melanin pigment and
its precursor are also produced by Vibrio cholerae, the causative agent of Asiatic Cholera, and an inhabitant of estuaries
throughout the United States. Outbreaks of cholera have
recently been reported (Morbidity and Mortality Weekly Reports,
Centers for Disease Control, 1978-1983) and V. cholerae is now
considered to be autochthonous to the estuarine environment
(Colwell et al. 1981). Without doubt, some fascinating
biological interactions have been discovered (Weiner et al.
1983).

The attachment process is important in the distribution of
bacteria in the aquatic environment. Cells adhere within
minutes after settling onto a surface (Belas and Colwell
1982a,b; Marshall 1976), and structures produced by the cells
are involved in attachment. Attachment to wood structures in
the marine environment is also rapid. Colonization occurs after
attachment of the cells, during which attachment structures are
synthesized, further binding the cells to the surface as well as
to each other.

Naphthalene-enriched creosote was examined as a means of
extending the life of wood dock pilings (Austin et al. 1979;
Colwell et al. 1980). Addition of up to 10 percent naphthalene
to creosote, for application to wood structures in the marine
environment, was related to delay in attachment. Eventually,
however, naphthalene-resistant organisms colonized the surfaces
of the wood, as with untreated wood.

Other materials have been examined, including stainless
steel, brass, and aluminum (Zachary et al. 1978). Parachute
material and epoxy adhesives, used by NASA for the booster unit
for the space shuttle, were also found to be readily colonized
(Zachary et al. 1978).

Fibril formation is interesting. Strain OWD-1, isolated
from old wooden pilings, readily colonizes surfaces, with
microcolony formation following the initial attachment event
(Colwell et al. 1980). Connecting fibrils are formed with
bacteria anchoring to each other as well as to the surface.
Subsequent to adhesion, agglomeration and colony formation
occurs. Cells that are singly located do not show the fibril
formation, but those that have been growing in aggregates show
extensive cross-formation.

The attachment and swarming of <u>Vibrio parahaemolyticus</u> on
surfaces has been described (Belas and Colwell 1982a,b). The
conclusion is that lateral flagella play a role in this attachment phenomenon (Belas and Colwell 1982a). The rate of attachment, using radioactively-labeled bacteria and substrates of
chitin, wood and glass, yielded very interesting data (Belas and
Colwell 1982b). Laterally-flagellated bacteria demonstrated a
Langmuir adsorption type of relationship, whereas polarly
flagellated bacteria showed proportional attachment occurring
wihin a very short period of time. Also, laterally flagellated
bacteria tended to exclude their polarly flagellated
counterparts (Belas and Colwell 1982b).

In a practical sense, very real problems arise from
biofouling. Microbial destruction of wood occurs at the water
line (Figure 5), where the degradation process proceeds further
with involvement of the marine isopod, <u>Limnoria tripunctata</u>
(Figure 6). The surface of <u>L. tripunctata</u> supports an extensive
bacterial flora (Figure 7). The relationship between these
surface bacteria and the <u>Limnoria</u> burrow, also layered with
bacteria (Figure 8), is that these bacteria may be invovled in
the degradation of wood (Colwell et al. 1980). In addition, the
gut microflora of <u>L. tripunctata</u> may be responsible for
resistance of the isopod to creosote (Zachary and Colwell
1979). Thus, bacteria play an interesting and important role in
association with this boring invertebrate.

Many of the bacteria observed by direct electron microscopy
cannot be recovered on laboratory media. Colonization and
growth of these microorganisms, however, can be studied using
acridine orange direct microscopy of cells that have been
stimulated by a nutrient, but simultaneously inhibited with
nalidixic acid (Kogure et al. 1979), an antimicrobic which acts
to prevent cell division but permits growth, i.e., elongation,
of cells. Perhaps with this direct viable technique, in
combination with molecular genetic techniques, the dynamics of
colonization can be further elucidated and the goal of selective
control of biofouling achieved.

Figure 5. Creosote-treated wooden dock pilings at the U.S. Naval Station, Roosevelt Roads, Puerto Rico. Extensive tunneling by L. tripunctata is visible at the water line (arrow) in Plate A. A closer view of the biofouling is shown in Plate B.

Figure 6. Scanning electron micrograph of the marine isopod
Limnoria tripunctata (50X).

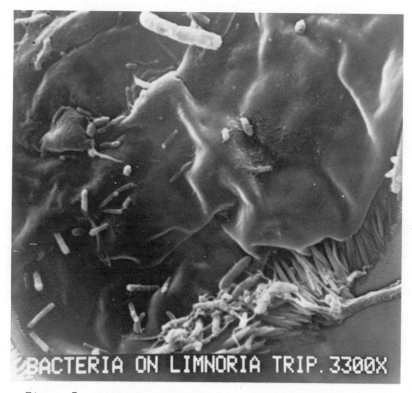

Figure 7. Scanning electron micrograph of L. tripunctata
showing bacteria attached to the surface (X3300).

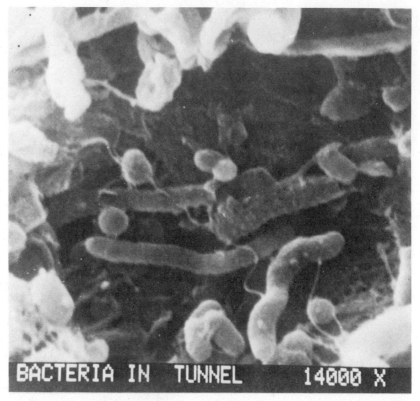

Figure 8. Scanning electron micrograph of a L. tripunctata
tunnel in creosote-treated wood. Note the extensive bacterial
colonization of the tunnel and morphological similarity to the
bacteria shown in Figure 7.

References

1. Belas, M.R. and R.R. Colwell, 1982a. J. Bacteriol., 150:956.

2. Belas, M.R. and R.R. Colwell, 1982b. J. Bacteriol.,
    151:1568.

3. Allen, D.A., B. Austin, and R.R. Colwell, 1977. Antimicrob.
    Agents Chemother., 12:545.

4. Berk, S.G. and R.R. Colwell, 1981. J. Exp. Mar. Biol.
    Ecol., 52:157.

5. Marshall, K.C., 1976. Interfaces in Microbial Ecology,
    Harvard University, Cambridge, MA.

6. Sochard, M.R., D.F. Wilson, B. Austin, and R.R. Colwell, 1979. Appl. Environ. Microbiol., 37:750.

7. Colwell, R.R., R.J. Seidler, J. Kaper, S.W. Joseph, S. Garges, H. Lockman, D. Maneval, H. Bradford, N. Roberts, E. Remmers, I. Huq, and A. Huq, 1981. Appl. Environ. Microbiol., 41:555.

8. Huq, A., E.B. Small, P.A. West, M.I. Huq, R. Rahman, and R.R. Colwell, 1983. Appl. Environ. Microbiol., 45:275.

9. Kaneko, T. and R.R. Colwell, 1973. J. Bacteriol., 113:24.

10. Weiner, R., A. Segall, P. Levon, P. West, S. Coon, D. Bonar, and R.R. Colwell, 1983. Third Int. Symp. Microbial Ecol., East Lansing, MI.

11. Austin, B., D.A. Allen, A. Zachary, M.R. Belas, and R.R. Colwell, 1979. Can. J. Microbiol., 25:447.

12. Colwell, R.R., M.R. Belas, A. Zachary, B. Austin, and D.A. Allen, 1980. Dev. Ind. Microbiol. 21:169.

13. Zachary, A., M.E. Taylor, F.E. Scott, and R.R. Colwell, 1978. Int. Biodeterio. Bull., 14:111.

14. Zachary, A. and R.R. Colwell, 1979. Nature, 282:716.

15. Kogure, K., U. Simidu, and N. Taga, 1979. Can. J. Microbiol., 25:415.

An Approach for Studying the Molecular Basis of Marine Microbial Fouling

Melvin I. Simon, Michael Silverman, Robert Belas, John Abelson, Dan Cohn, Alan Mileham, Richard Ogden, and Marcia Hilmen

Agouron Institute
LaJolla, California

Abstract

Primary fouling of marine surfaces involves the adsorption of bacteria and their adhesion to the surface. These are basic properties of the organism controlled by specific genes and complex regulatory systems. In order to understand the processes that regulate bacterial adhesion, genetic studies have been undertaken. These studies involve methods to identify and modify the genes that encode substances required for adhesion. The genes themselves can then be isolated and characterized and their products studied. In order to develop these approaches in marine bacteria, the bioluminescence system was used. Approaches designed to characterize and isolate the genetic correlates of this system are described. These same methods will be used to study adhesion. It is hoped that a molecular understanding of the process will provide clues to developing procedures to control adhesion in the marine environment.

Bacteria play a central role in a number of processes in the ocean. They are important components of the food chain and are required for nutrient turnover. They are also responsible for the initial stages of some of the processes involved in marine fouling and biodeterioration. There is a growing body of evidence to indicate that the ability of bacteria to adhere to surfaces plays a critical role in determining their function, both in the marine and terrestrial environments (Savage 1977). Recent studies describe the specific interactions of bacteria with a number of different kinds of surfaces (Bitton and

233

Marshall 1980). For example, microbial infections of many sorts
apparently involve specific interactions of binding proteins on
the surface of bacteria with components on the surface layer of
the animal cell. Associations between bacteria and plant root
and leaf surfaces have been shown to be important in the
development of nitrogen fixing systems and in the regulation of
plant diseases. There is evidence indicating that marine
bacteria exist adsorbed onto the surface of particles in a
sessile form. In fact, the ability of the organisms to adsorb
to clean surfaces is the basis for one method used for sampling
marine bacteria (Zobell 1943). The remarkable adhesive
properties of these organisms are also the source of a variety
of problems. Almost every substance introduced into the marine
environment is subject to biodeterioration and fouling. Generally, it appears that after an initial encounter, bacteria are
able to form specific attachments that involve the interaction
of bacterial products with the surface introduced into the
ocean. This is the start of a complicated process which results
in the adhesion of a variety of other organisms and coating of
the surface. Thus, adhesion, which is a basic cause of marine
fouling, is also a basic property of all bacterial cells, and it
is manifested both in the marine and terrestrial environments.

In order to interact with any material, the bacterial cell
must first find the surface, then attach, adhere, and elaborate
substances that foster adhesion. These substances act as substrates for other organisms. These functions, of course, are
all genetically controlled. The products of the genes for the
chemotaxis and motility systems may allow bacteria to find the
surface. Genes that code for a variety of organelles and
surface components, including the cell wall and cell surface
polysaccaharides and polypeptides, are required for adhesion.
Other genes are involved in elaborating the capsular polysaccharides which form the matrix in which bacteria are maintained
on a variety of surfaces (Fletcher and Floodgate 1973). There
are also genes which allow bacteria to resist heavy metals and
other poisons so that they are able to persist under adverse
conditions. The regulation of expression of all of these
properties is genetically controlled. Thus, for example, two
different kinds of flagella may be synthesized, depending upon
the conditions under which the bacteria are grown and the
surfaces to which they are exposed (Belas and Colwell 1982).
These kinds of regulatory mechanisms are essential in order to
allow the organism to adapt and perform optimally in a variety
of situations. By defining these genes and processes, we can
begin to understand in molecular terms the sequence of steps
that allows organisms to interact with an enormous number of
different surfaces.

Fouling occurs relatively rapidly and leads to tremendous
losses in the performance of ships and other marine equipment.
In the past, this problem has been dealt with by the preparation
of coatings that leak toxic substances into the environment and

prevent fouling, presumably by inhibiting bacterial growth and metabolism. Unfortunately, paints containing heavy metals and other coatings have limited lifetimes, and mechanical cleaning or oxidation is eventually required at great cost in terms of energy expenditures and loss of operational performance.

At the Agouron Institute we have approached the question of fouling from a fundamental biological point of view. We believe that if we could understand specific molecular mechanisms involved in the basic process of bacterial adhesion, we could determine how to intervene in the processes so that preventive methods could eventually be developed. We will examine simple model systems in attempts to understand general mechanisms that may be common to a variety of organisms. It is now possible to employ genetic approaches to allow systematic exploration of the biochemical pathways involved in complicated processes in heretofore unmanageable bacteria.

The introduction of recombinant DNA technology in the past three or four years has brought about a revolution in biological research comparable to the changes experienced in physics as a result of technological advances thirty or forty years ago. These new techniques allow us to apply genetic analysis to a variety of organisms and to understand the molecular basis of a biological function. Thus, it is possible to obtain a complete catalogue of all the genes that specify the functions of an organism. For a bacterium, such a collection may be composed of about 2,000 clones (Blattner et al. 1977), each of which contains part of the genetic complement of the organism. One can then select from such a collection a subset of genes that specifies the functions of interest. Each subsystem can then be analyzed to determine the biochemical and structural steps involved in its formation, regulation, and function. This kind of technology is finding application in many areas. We have been using this approach to study motility and chemotaxis (Boyd and Simon 1980) as well as systems that involve direct changes in gene structure that control gene expression (Simon et al. 1980). We believe that these kinds of regulatory mechanisms may be ubiquitous and account for a great deal of the observed ability of bacteria to adapt to a wide variety of environments.

In order to develop a genetic system that can be used with marine bacteria, we have decided to work initially with the Gram-negative organisms, Vibrio harveyi and Vibrio parahaemolyticus. Techniques applied to the Vibrio species will be useful for other Gram-negative bacteria and in a modified way for some Gram-positive bacteria. We have obtained a complete library of the genetic material of V. harveyi. The genes that correspond to a series of subsystems that are of particular interest with respect to the properties of these organisms in the marine environment are being isolated from the library. We have identified the genes for bioluminescence and are currently attempting to identify the DNA corresponding to the genes involved in cell motility, adhesion, chemotaxis, heavy metal

resistance, the formation of capsular material, and cell surface
antigens. Some of these functions are determined by only a few
genes, while others involve the complex interaction of many
genes. In previous work in our laboratory, we have found that
this kind of analysis can in fact be done even in relatively
complex systems. The genes can be isolated, their products
determined, and the mechanisms involved in turning them on and
off can be studied (Silverman 1980).

At present, we know so little about the biochemistry and
genetics involved in adhesion and fouling that it is difficult
to predict the specific approaches that will emerge. For
example, we may determine that it is necessary to know more
about the nature of specific gene products. It is possible to
use recent techniques to introduce known changes into the gene
by resynthesizing genes with altered nucleotide sequences.
These human-made mutations can then be reinserted into the
bacterium to determine how they affect specific behavior. A
great many other novel approaches are possible depending upon
initial findings.

Gene Isolation

A balanced genetic approach requires both the isolation of
bacterial genes on autonomously replicating vehicles and the
development of tests and probes to recognize the particular
genes of interest in the gene bank. We encountered no severe
difficulties in constructing banks of DNA fragments from various
Vibrio bacteria. Once the DNA was purified, cutting the Vibrio
DNA with specific restriction endonucleases and inserting the
fragments into DNA cloning vehicles was straightforward.
Initially, a variety of bacteriophage lambda vehicles were used
to make gene banks. These vectors could accept up to 20kb of
DNA and accomodate DNA fragments made with a variety of restriction endonucleases (Blattner et al. 1977). The gene banks
consisting of hybrid bacteriophage vectors were maintained
either as infective virus or in the latent (lysogenic) state in
Escherichia coli. Recombinant clones can be distinguished from
vector molecules which do not contain Vibrio DNA inserts by
testing for specific indicator functions incorporated into the
genetic structure of the cloning vehicles. Appropriate tests
exist for screening thousands of recombinant clones for a given
gene in a few days' time. Provided that the restriction enzyme
used to fragment the Vibrio genomic DNA did not interrupt the
gene of interest, the gene banks made with bacteriophage vectors
should contain the gene at a frequency of approximately one per
2000 hybrid clones.

Our first attempt to isolate specific genes was directed to
the lux genes which encode the protein components responsible
for bacterial luminescence. The central components are the
alpha and beta proteins encoded by two genes (luxA and luxB) on
the Vibrio harveyi chromosome. We chose this system because it
was well characterized biochemically and we could use the known

amino acid sequence of parts of the luciferase gene (Belas et al. 1982) in order to design an oligonucleotide probe. A mixture of eight different oligonucleotides, each seventeen nucleotides long, was synthesized in order to account for the degeneracy of the DNA code.

Thus, we were certain that this oligonucleotide mixture contained one DNA sequence corresponding to a short length of the luxA gene (only 5-2/3 amino acids of coding capacity). The oligonucleotide was radioactively end-labelled with ^{32}P phosphate and used to identify the hybrid clone which contained the luxA gene. Using DNA hybridization, which indicated the presence of complementary DNA sequences in a lambda vector gene bank, a positive signal was obtained for one lambda clone after examining about 5,000 hybrid lambda clones. In practice, this involved the following steps: (1) making a DNA "imprint" of hybrid lambda bacteriophage growing in a large two dimensional display; (2) allowing the radioactive probe to react with the imprint; (3) locating positive radioactive signals on the imprint; and (4) referring back to the master file to isolate the positive clone. Upon detailed analysis of its structure by DNA sequencing methods the positive clone proved to contain the entire luxA gene and a short segment of the beginning of the luxB gene. Thus, from a small amount of DNA sequence information deduced from the protein amino acid sequence, the entire gene encoding that protein could be isolated. The complete nucleotide sequence of these genes is currently being determined.

Transposon Mutagenesis

While the approach we have outlined worked with lux where we had the benefit of a great deal of biochemistry and genetics defining the components of the system (Belas et al. 1982), a more general approach will be required to get at adhesion where the system is undefined. Mutants defective in the expression of adhesion functions will be of fundamental importance. These mutants will define adhesion factors and will be used to identify the cloned adhesion genes. Mutants obtained from conventional chemical mutagenesis are unsatisfactory in this situation because: (1) these mutants have a broad spectrum of phenotypes; (2) the site of the mutation is very difficult to locate because of the lack of hundreds of standard markers for linkage studies in these wild organisms; (3) the mutations are difficult to transfer and recombine because few if any conjugative or transductional systems exist; and (4) there is no selectable phenotype associated with the mutation.

Transposons (Kleckner 1981) are mobile genetic elements and are very powerful biological mutagenic agents which produce mutations with characteristics particularly beneficial for genetic analysis. Transposons are DNA elements which replicate as part of the larger DNA molecule in which they reside, but have the ability to transpose a copy of themselves or "hop" to other locations. These elements are of the order of thousands

of base pairs (kbp) of DNA, and when inserted into a new location, for example into a gene encoding an adhesion protein or
luciferase protein, the continuity of that region is interrupted. This disrupts the decoding of genetic information, and
prevents expression of any genes "downstream", i.e., other genes
read in the same operon. Thus, transposon mutants have
completely lost function. This null phenotype makes interpre-
tation of the nature of the effect of the mutation simpler.
Also, the mutation is produced by a large structural change, the
insertion of a piece of extraneous DNA, which makes it possible
to determine the location of the mutation. Furthermore, a
highly useful characteristic of the transposon is that this
mobile element carries a gene which encodes one of a variety of
drug resistances. Thus, the mutagen contains a selectable trait
which is, for the most part, inseparable from the mutant
phenotype.

Drs. Belas and Silverman have applied this technology to
marine Vibrio bacteria and have demonstrated the usefulness of
this genetic tool (Belas and Silverman in press). A Vibrio
harveyi strain (BB7) was found to be amenable to this kind of
genetic approach (Figure 1). Transposon Tn5-132, a 5 kb element
encoding tetracycline resistance, was transferred to this strain
using the coliphage P1. The Tn5-132 element resided on the P1
genome and was injected upon infection of BB7 into the cell.
Since the P1 genome does not replicate in this organism, the
transposon must "hop" or transpose to the V. harveyi chromosome
to be stably inherited. BB7 bacteria infected with the P1 phage
carrying the transposon were isolated and analyzed to determine
if the transposon had indeed inserted in a lux gene. We were
able to show that the Lux⁻ phenotype was the result of insertion of the transposon DNA into the lux genes encoding the alpha
and beta subunits of bacterial luciferase (Figures 2 and 3).

Once a transposon mutant is isolated, the target gene is
genetically linked to the transposon, and the mutant phenotype
is constantly associated with the transposon's drug resistance
phenotype. Having isolated Lux⁻ transposon mutants, it is
simple in principle to clone the lux gene into which the
transposon has inserted. By appropriate selection of restriction endonucleases, a DNA fragment containing the lux gene can
be cloned into a vehicle and identified by selection for the
transposon encoded drug resistance -- in this case tetracycline
resistance. We have isolated the region of Vibrio harveyi DNA
encoding the lux genes by this method. Two clones were
isolated, each with a Tn5-132 transposon located at different
sites in the lux gene cluster, (Figure 3). These two genes are
closely linked on the Vibrio harveyi chromosome and both genes
are on the aforementioned clones. Of course, both clones are
defective for lux due to Tn5-132 insertion. To reconstruct a
functional lux cluster, we rely on a classical in vivo
approach. That is, we use the cell's DNA recombination system

TRANSPOSON MUTAGENISIS

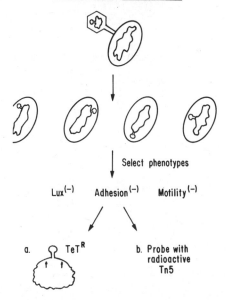

Select phenotypes

Lux$^{(-)}$ Adhesion$^{(-)}$ Motility$^{(-)}$

a.   TeT^R b. Probe with
                              radioactive
                              Tn5

Figure 1: Strategy for transposon mutagenesis. Coliphage Pl, which contains transposon Tn5-132 within its genome, was used to infect Vibrio strains. Pl DNA is injected, but does not replicate in Vibrio cells. Therefore transposon Tn5-132 can persist only if it inserts into the Vibrio chromosome. Mutants which contain Tn5-132 insertions were tetracycline resistant, since the transposon encodes tetracycline resistance. A variety of mutant phenotypes could be selected from a collection of mutants generated by random transposon insertion. The mutated gene could then be recovered by cloning in E. coli and selecting for cloned DNA fragments carrying the tetracycline resistance marker (a). The location of the transposon insertion within its target gene could be determined by hybridization with a radioactively labeled DNA probe specific for Tn5-132 (b).

to stitch a complete copy out of two copies defective in different parts.

Both lux gene clones were made to cohabit the same E. coli cell by cloning each onto a different, but compatible, plasmid vehicle and introducing both into the same bacterium. Then the bacterium's natural recombination system catalyzed DNA exchanges between homologous segments of the two plasmids. By sorting through DNA molecules isolated from this population of bacteria, we found a few plasmids in which a natural crossover event had reconstructed the non-defective complement of lux genes (Belas et al. 1982). Interestingly, when the proper co-factors were

Vibrio harveyi lux genes

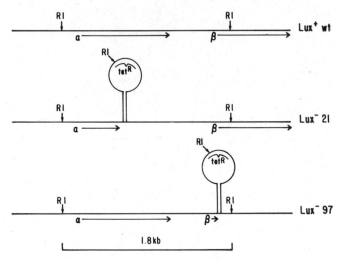

Figure 2: Location of Tn5-132 insertions by DNA hybridization. Vibrio harveyi DNA was cleaved into small fragments with restriction enzymes and separated according to size on agarose gels. These gels are shown in the figure. DNA fragments were transferred onto nitrocellulose paper and hybridized with the ^{32}P-labelled DNA probes specific for lux genes or the Tn5-132 transposon. Positive hybridization was detected by autoradiography on X-ray film. Hybridization with the radioactive probes occurred only with particular DNA restriction fragments among the thousands of fragments present. When DNA fragments from the wild-type Vibrio harveyi were probed with labelled lux gene DNA, particular fragments hybridized. Hybridization with DNA from the Lux⁻ transposon mutants indicated that the lux gene region had been altered by the insertion of transposon DNA. Using a variety of restriction enzymes the precise location of insertions in the mutants Lux⁻21 and Lux⁻97 could be determined. Restriction enzymes used were: EcoR1 (E), BamH1 (B), and Hind III (H). Sizes of fragments in kb are shown at right.

supplied, the E. coli which contained this hybrid plasmid showed bioluminescence, demonstrating this property for the first time in a bacterium unrelated to V. harveyi. The essential element in this procedure was the transposon mutagenesis. This method should be entirely general and should apply to genes which encode adhesive factors as well.

We have thus far succeeded in applying a number of the standard genetic and recombinant DNA techniques to the marine vibrios and particularly to the bioluminescence system derived

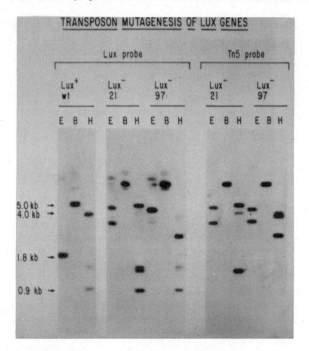

Figure 3:   Location of transposon insertions in the lux gene
region of Vibrio harveyi.   Hybridization information from Figure
2 was used to position the transposon Tn5-132 insertions in
Lux⁻21 and Lux⁻97 mutants.   Transposon Tn5-132 inserted in
the region N-terminal to the luxA gene in the lux-21 mutant and
near the N-terminal part of the luxB gene in the Lux⁻97 mutant.
EcoRl (Rl) restriction sites are shown.   Arrows denote the
coding sequences and direction of transcription.

from Vibrio harveyi.   Our initial approach has shown that all of
these techniques are feasible and that they can be used in a
general way to approach many of the properties of marine
bacteria.   We are hopeful therefore that we will be able to
approach the adhesion problem in all of its complexities and
analyze the genetic activities required for marine microbial
adhesion.   This is not to say that a simple clear pattern will
immediately emerge, nor that there is any guarantee that this
kind of work will lead to simple, technical solutions to the
problem.   However, approaching the systems from the level of
gene activities allows an analytical approach to the study of
the behavior of these organisms and their relationship to the
fouling process.
    We are convinced that knowing more about this system will,
in the long run, contribute to ways of intervening and dealing
with the problems that result from microfouling.   This kind of
approach to the problem does not promise a quick fix.   It does,

however, suggest that we will not only understand the factors involved in microbial adhesion, but will also be able to manipulate these factors at a genetic and biochemical level. We take it as axiomatic that an understanding of the process of microbial adhesion will put us in a better position to deal with the problem of microfouling than if we approach the same problem from ignorance.

Acknowledgements

This work was carried out at the Agouron Institute, LaJolla, California, and was supported by a contract from the United States Navy, Office of Naval Research.

References

1. Belas, M. and R. Colwell, 1982. J. Bacteriol., 150:956.

2. Belas, R., A. Mileham, D. Cohn, M. Hilmen, M. Simon and M. Silverman, 1982. Science, in press.

3. Belas, R. and M. Silverman, submitted for publication.

4. Bitton, G. and K. Marshall, Eds., 1980. Adsorbtion of Microorganisms to Surfaces, John Wiley & Sons, New York.

5. Blattner, F., B. Williams, A. Blechl, K. Denniston-Thompson, H. Faber, L. Furlong, D. Grunwold, D. Krefer, D. Moore, J. Schumm, E. Sheldon and O. Smithies, 1977. Science, 196:161-169.

6. Boyd, A. and M. Simon, 1980. Ann. Rev. Physiol., p. 501-517.

7. Cohn, D., R. Ogden, J. Abelson, T. Baldwin, K. Nealson, M. Simon, and A. Mileham, 1982. Proc. Natl. Acad. Sci., U.S.A., in press.

8. Fletcher, M. and G. Floodgate, 1973. J. Gen. Microbiol., 74:325.

9. Kleckner, N., 1981. Ann. Rev. Genetics, 15:341.

10. Savage, D.C., 1977. Ann. Rev. Microbiol., 31:107.

11. Silverman, M., 1980. Quart. Rev. Biol.

12. Simon, M, J. Zieg, M. Silverman, G. Mandel, and R. Doolittle, 1980. Science, 209:1370.

13. Zobell, C.E., 1943. J. Bacteriol., 46:39.

Biofouling in Freshwater Cooling Systems

Denise S. Richardson

The Mogul Corporation
Chagrin Falls, Ohio

Abstract

Industrial cooling systems that use once-through or recirculating water are conducive to macro- and microbiological
fouling. Certain bacteria corrode piping and metal structures
and form troublesome slime on surfaces; shellfish, mussels, and
barnacles block pipes. There are mechanical and chemical ways
of controlling biofouling, but there is a need for research to
investigate the effects of corrosion-inhibiting films, biocides,
and chemical agents on biofouling and modification of the
environment to inhibit biofouling.

Industrial systems may use once-through or, more commonly,
recirculating fresh water, once-through seawater, or brackish
water for cooling. These systems are subject to both micro- and
macrobiological fouling. Open evaporative recirculating cooling
systems scrub windborn dust and organisms from the air, concentrate nutrients, and provide a warm environment for microbial
growth.

A simplified open evaporative recirculating system is
presented schematically in Figure 1. Warm water enters the top
of a cooling tower (A) and flows over the deck and through
distribution holes (B). The water forms drops as it hits the
tower fill (C) and loses heat through evaporation. Fill
materials are commonly wood, PVC or asbestos. Cooling water
collects in the tower basin (D) and is pumped (E) through the
system, eventually reaching the heat exchanger (F) where it can

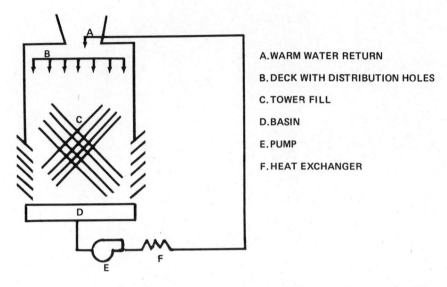

A.WARM WATER RETURN

B.DECK WITH DISTRIBUTION HOLES

C.TOWER FILL

D.BASIN

E.PUMP

F.HEAT EXCHANGER

Figure 1: Simplified Open Evaporative Recirculating Cooling System

absorb more heat.   Areas in the system which are subject to biofouling include the tower structure, often constructed of wood and piping.

Hazards of Biofouling
The tower deck and fill, if open to air and light, provide an ideal place for algae to grow. Mats of algae may form on the deck, plugging distribution holes and interfering with water flow. Algae on the tower fill can prevent droplet formation and decrease the rate of heat removal. McCoy (1980) lists Oscillatoria, Chlorella, and Ulothrix as the three most prevalent troublesome genera. Cellulolytic fungi of the class Fungi imperfecti attack the wooden portions of the cooling tower, causing surface decay also known as soft rot. Cellulolytic basidiomycetes may cause non-visible internal decay or pocket rot (McCoy 1980 and Smith and Muia 1973). The results of this type of biofouling can be a dangerous and unnoticed weakening of the wooden structure.
Bacteria may cause corrosion of piping and costly material deterioration. Some of the most troublesome bacterial types encountered in cooling systems are Desulfovibrio desulfuricans and other sulfate reducing bacteria. Strictly anaerobes, these bacteria utilize sulfate ($SO^4=$) and produce the objectionable H_2S gas. H_2S reacts with ferrous based metals to form black FeS deposits. Metal corroded by sulfate reducing bacteria tends to be pitted. Corrosion can be severe enough to perforate the

metal completely. Postgate (1979) has published a monograph on
sulfate reducing bacteria, their classification, ecology, and
economic importance.

Bacteria which form slime (Gallionella, Pseudomonas, and
others) may increase frictional resistance, decrease heat
transfer, and increase pressure drop through piping. The
increased pressure drop due to biofilm may be greater than that
expected due to the simple reduction of cross sectional area in
the pipe (Norrman, Characklis, and Bryers 1977). The results of
biofilm formation are increased costs for equipment and energy,
for which the consumer must ultimately pay. Atwood (1980)
estimated that upwards of $100 million per year could be saved
by turbine backpressure reduction of 0.1 in. Hg absolute, based
on the utilities' situation in 1977.

Shellfish, mussels, and barnacles cause mechanical blockage
of pipes. Systems may need periodic shutdowns so that shells
can be mechanically scraped from accessible surfaces. Acrolein,
chlorine or heat have been used to control macrobiological
fouling (Krishna Swamy 1975 and Hamm, 1982). Acrolein, however,
is strictly regulated by the U.S. Environmental Protection
Agency (EPA) and is a lachrymator.

Controlling Biofouling

Biofouling in industrial systems is generally controlled by
mechanical or chemical means. Various brushes or inserts with
abrasive surfaces may move through the piping, scraping away
deposits. Mechanical cleaners are available from a number of
commercial suppliers. Chemicals added to the system are normally the most useful method to control the growth of microorganisms. These pesticides must be registered with the EPA.
Some commonly used biocides are chlorine (or chlorine donors
such as calcium hypochlorite), methylene bisthiocyanate,
carbamates, dibromonitrilopropionamide, isothiazolins, and
quaternary ammonium compounds. Selection of the best biocide
depends on the system's water chemistry and compatibility with
corrosion inhibitors.

Pesticides, added to control the growth of organisms in the
recirculating water, may not be as effective on organisms
imbedded in slime. Extracellular polysaccharides may serve to
protect the cell from harmful molecules. Characklis, Truleah,
and Stathopoulos (1980) have shown that chlorine may react with
biofilm and strip it from the surface. Tighter regulations on
the discharge of chlorine and concern over production of tri-
halomethanes and chloramines demand judicious and responsible
application of this biocide by the user. Decreasing resources,
increasing costs of water and energy, and concern for the
environment require that all pesticides be applied in the safest
and most effective manner. Discharge of pesticides is also
regulated by state and federal agencies.

Areas of interst in biofouling, which both industry and academia may need to investigate include:

1. Modification of surfaces to inhibit both corrosion and bacterial attachment
2. The effect of corrosion inhibiting films on biofouling
3. The effect of biofilm thickness on the activity of biocides: Is there a critical film thickness above which a biocide is not effective on the organisms embedded in the film?
4. A search for chemical agents that inhibit attachment
5. A way to modify the cooling tower environment so that bacteria preferentially grow in a planktonic rather than sessile form, making them vulnerable to biocides added to the recirculating water.

References

1. Atwood, K.E., 1980. Cost impact of condenser fouling. In J.F. Garey et al, Eds., Condenser Biofouling Control, Ann Arbor Science, Ann Arbor, Michigan, pp. 43-45.

2. Characklis, W.G., M.G. Truleah and N. Stathopoulos, 1980. Fundamental considerations in biofouling control, Cooling Tower Institute Annual Meeting, Houston, Texas.

3. Hamm, P.J., 1982. What are we doing about the Asiatic clam? Power, 126:25-28.

4. Krishna Swamy, M.S., 1975. Controlling marine growth in cooling-water tunnels. Chem. Eng., 82:140.

5. McCoy, J.W., 1980. Microbiology of Cooling Water, Chemical Publishing Co., Inc., New York, p. 28-29.

6. Norrman, G., W.G. Characklis and J.D. Bryers, 1977. Control of microbial fouling in circular tubes with chlorine, Dev. Ind. Microbiol., 18:581-590.

7. Postgate, J.R., 1979. The Sulphate Reducing Bacteria, Cambridge University Press, Cambridge.

8. Smith, A.L. and A.R. Muia, 1973. Identify and control microbiological organisms in cooling water systems. Power, 117:16.

IV. MARINE POLLUTION CONTROL

Biotechnology and Environmental Pollution

Scott T. Kellogg

Bethesda Research Laboratories
Gaithersburg, Maryland

Abstract

The release of xenobiotic halogenated hydrocarbons, inorganics, and heavy metals into the environment has resulted in widespread environmental and human contamination. Xenobiotic halogenated hydrocarbons originate as pesticides or industrial chemicals and are both persistent and toxic as a result of the halogen-carbon bond which is not easily broken, since the required enzymes are rare in microbial populations. Microbial populations infrequently face evolutionary pressures to evolve such enzymes due to low levels of exposure (ppm) to particular chemicals coupled with a variable distribution of both the xenobiotic compounds and the microorganisms themselves. One possible approach to eliminating these toxic chemical pollutants is through genetic engineering of specific bacteria to equip them with the genes for the necessary degradative enzymes. New genetic techniques, such as plasmid-assisted molecular breeding (PAMB) and state-of-the-art genetic engineering, are presented here along with specific examples of newly engineered degradative bacteria for use in cleaning up both marine and terrestrial pollution.

Before examining the possible applications of biotechnology to environmental pollution, definitions of both biotechnology and pollution are required. Biotechnology is defined here as the application of the latest advances in molecular biology, genetics and cell biology to solutions of important problems. Pollution is considered a biologically persistent and atypical

intrusion into the environment. Biological persistence is cited
since chemical persistence may not directly relate to environmental impact. For example, highly persistent compounds may
have minimal impact whereas short half-life compounds may have
maximal impact. Chemical persistence does not always follow
loss of toxicity. Any atypical eutrophication (usually caused
by man) is considered to be an environmental pollution problem.

Specific examples of industrial chemical pollution in marine
and terrestrial ecosystems might be polychlorinated biphenyls
(PCBs), pulp mill wastes, and fire retardants (polybrominated
biphenyls, PBBs). Some examples of heavy metal and inorganic
pollution might be mercury, cadmium, iron, nitrate and phosphate. The latter two inorganic compounds most commonly
originate from fertilizer runoff or earthmoving operations.
Lastly, petroleum and its derived products (crude oil, diesel,
fuel oils, etc.) contribute to environmental pollution,
especially in marine systems.

The scope of this paper will be confined to possible
applications in marine estuarine and neritic systems as well as
terrestrial systems. Marine pelagic pollution will not be
covered due to the scale of the problem.

History of Environmental Pollution Control

The entire area of environmental pollution is far too broad
to cover here; however, one specific type of pollution needs to
be discussed, since it was responsible for the first research
into possible genetic engineering in environmental pollution.

Traditionally oil spills, as well as most pollutants, have
been controlled (not eliminated) by dredging, adsorption
methods, and physical containment. Specific research in biotechnology was first directed towards the treatment of oil
spills. A.M. Chakrabarty, while working at the General Electric
Co., created a Pseudomonas strain, via plasmid manipulation,
that had the metabolic capability to degrade crude oil (Friello
et al. 1976). The intent was to use this genetically created
bacterium in areas where oil spills had been caused by oil
tanker accidents, offshore drilling leaks, etc. Table I
illustrates the genetic engineering steps involved in creating
this strain as well as its composite phenotype. In order to
protect the commercial value of such research, a successful
attempt was made to patent this bacterium (Chakrabarty vs.
Diamond 1980). The Supreme Court ruling in this case was
important since it was the first time that an organism created
with plasmids -- one of the basic tools of genetic engineering
-- was covered by such legislation. However, this initial legal
and scientific success was never commercially tested, due mostly
to the low economic incentive inherent in such a test, the
infrequency of oil spills, and their relatively rapid natural
degradation.

Although petroleum pollution is significant in many locations, especially marine, the problem of toxic and persistent

Table I. Construction of multi-plasmid strain from different naturally occurring mono-plasmid parents (Friello et al. 1976).

Step	Donor Organism	Plasmid	Recipient Organism	Resulting Exconjugant Phenotype[a]
(1)	P. putida AC59	CAM-OCT	P. aeruginosa	$Cam^+ Oct^+$
(2)	P. putida AC137	XYL	$Cam^+ Oct^+$ P. aeruginosa	$Cam^+ Oct^+ Xyl^+$
(3)	P. putida PpG7	NAH	$Cam^+ Oct^+ Xyl^+$ P. aeruginosa	$Cam^+ Oct^+ Xyl^+ Nah^+$

[a] Abbreviations: CAM, camphor; OCT, octane; XYL, p- or m-xylene; NAH, naphthalene. Cam^+ indicates that the recipient organism possesses the ability to degrade camphor, etc.

chemicals in the environment poses a danger not only to the
environment but also directly to human health. The Love Canal
episode, Hudson River and Waukegan Harbor PCB contaminations,
PBB poisoning throughout Michigan, and the Seveso dioxin inci-
dent are only a few of the well known examples of halocarbon
pollution. It was the recognition that such dangerous episodes
could recur that prompted the initiation of genetic research to
create specific microorganisms capable of completely metabo-
lizing the compounds responsible for these disasters.

Genetic Approaches to Environmental Pollution

The U.S. Environmental Protection Agency has delineated over
one hundred priority chemical pollutants, two-thirds of which
are halogenated hydrocarbons. This paper focuses on compounds
harmful to humans. Halogenation of hydrocarbons lends chemical
and physical stability to these organic compounds by making them
temperature and electricity resistant and thus poorly autode-
gradable; halogenated hydrocarbons are also environmentally
persistent, a highly useful characteristic in pesticides since
only few applications are needed to eliminate selected plant and
insect pests. However, since these synthetic compounds are
xenobiotic (natural halo-organics are rare) the microbial populations responsible for mineralization do not possess the
metabolic capability to degrade them. Not only are degradative
enzymes required, but a mechanism must operate to deal with the
original toxicity, and with the possible toxicities of metabolites.

The major problem in genetically engineering the necessary
bacterial strain was that no a priori knowledge was available
for inserting the required enzymes and genes. In fact, not even
the dehalogenation gene(s) have yet been isolated, let alone
manipulated via recombinant DNA technology. Thus a technique
was developed which rested upon the existing degradative gene
library available (on plasmids) as well as potential wild type
degradative genes. This technique has been called Plasmid-
Assisted Molecular Breeding (PAMB) (Kellogg et al. 1981). The
technique is illustrated in Figure 1. Basically PAMB consists
of using continuous culture (in a chemostat) of strains
containing several molecular breeding plasmids, e.g., TOL, SAL,
pAC25, etc., as well as wild type potentially degradative
bacterial strains. The original breeding plasmids are supplied
initially with their respective substrates, but over time the
parental substrate concentrations are lowered and the new
substrate (in this case a toxic compound) concentration level is
gradually raised. This puts immense selective pressure on these
transmissible plasmids, not only to migrate (most probably via
transconjugation), but also to form the necessary recombinants
via one or more genetic mechanisms, e.g., transposon rearrangement, plasmid and/or chromosome recombination, deletion, gene
duplication, etc.

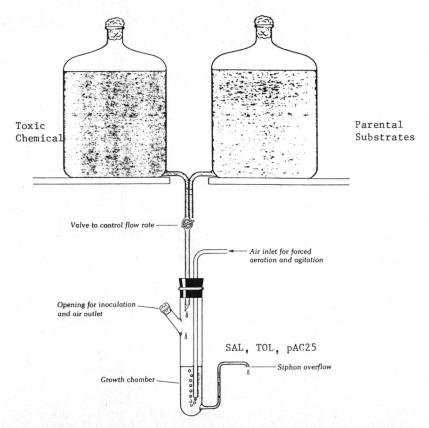

Toxic
Chemical

Parental
Substrates

Valve to control flow rate

Air inlet for forced
aeration and agitation

Opening for inoculation
and air outlet

SAL, TOL, pAC25

Siphon overflow

Growth chamber

Figure 1. Schematic of Plasmid-Assisted Molecular Breeding
(PAMB) technique (see text for specific details).

Current Research

Specific chemicals that have been studied with PAMB are
chlorobenzoates, chlorophenols, chlorophenoxy herbicides, PCBs,
and dioxins. The powerful defoliant Agent Orange (Herbicide
Orange) is composed of a 50:50 combination of 2,4-dichloro-
phenoxyacetic acid (2,4-D) and 2,4,5-trichlorophenoxyacetic acid
(2,4,5-T). Results of degradation experiments in a soil matrix
(see below) with a 2,4,5-T PAMB-generated culture are summarized
in Table II. These results with resting cells (grown on appropriate substrate, washed in buffer repeatedly, and resuspended)
in soil or in buffer are similar, however ca. 10 fold faster in
buffer alone (Kellogg et al. 1981). Halogen release was
demonstrated by ion specific electrode chemistry, while overall
metabolism was seen in scanning spectrophotometry and gas
chromatography. Very rapid degradation was seen as well as

Table II. Metabolism of 2,4,5-T as sole carbon source in soil.[a]

Sample	Incubation Time (hr.)	2,4,5-T Degradation (%) Spectrophotometric	Gas Chromatographic	Chloride Release (%)
Control	0.25	0.1	0.1	0.1
Test		6.5	5.2	7.4
Control	48	0.1	0.1	0.1
Test		68.8	82.5	58.6
Control	96	0.1	0.1	0.1
Test		89.6	97.1	99.4

a 2.5×10^8 cells/g soil. Controls were uninoculated. Results were similar for concentrations of 100-1000 ppm 2,4,5-T.

activity over a large range of concentrations (0-3000 mg/1).

Within the soil the chlorophenoxy herbicides vary in half lives, ranging from 25 days for 2,4-D to 106 days for 2,4,5-T at a concentration of 1 ppm (McCall et al. 1981). Higher concentrations, for example 1000 ppm, are expected to give half lives of greater than 1-3 years. It also should be pointed out that 2,4-D degradation can be plasmid mediated (Pemberton et al. 1979). The question of removal of a toxic compound from soil was examined by simply inoculating soil with resting cell suspensions. It was shown that greater than 99 percent of 2,4,5-T was rapidly removed even at high concentrations, e.g., 1000 ppm in 3 days at 30°C (Table II).

The next question was whether plants could survive in soils after chemical removal since unknown toxic materials still might remain. At concentrations from ca. 10 ppm to 2000 ppm a single soil inoculation sufficed to remove 99-100 percent of the chemical in 1-2 weeks; however, high concentrations required weekly resting cell inoculations to achieve similar removals. Concentrations as high as 20,000 ppm were reduced 99.5 percent in only 6 weeks, which should be compared with a normal half-life of many years.

The degradation pathway of this 2,4,5-T strain was studied via scanning spectrophotometry with authentic metabolites, as well as with gas chromatography and combined gas chromatography/mass spectrometry/data system analysis (GC/MS/DS). Thus far the first metabolic step has been confirmed as conversion to 2,4,5-trichlorophenol which agrees with the degradation pathway reported earlier (Loos, 1979; Que Hee and Sutherland 1981). However, other metabolites seen in GC/MS/DS analysis do not correspond to reported pathway products, and further analyses are proceeding.

Lastly, the nature of substrate specificity was examined by respiration studies employing resting cells and polarography. Results can be seen in Table III. Unexpectedly, there was a broad range of enzyme activity which is contradictory to previously reported high degrees of halocarbon specificity (Dorn et al. 1974; Dorn and Knackmuss 1978; Reineke and Knackmuss 1978; Hartmann et al. 1979). As expected, the first metabolite 2,4,5-trichlorophenol has a respiration rate similar to the parent compound; however, the dihalogen 2,4-D shows an almost equal rate, and most interestingly the nonhalogenated parent compound, phenoxyacetic acid, shows little or no respiration, indicating a possible requirement for halogens. Previous reports generally indicate that increased halogen presence reduces degradation rate, e.g., with PCBs (Tucker et al. 1975; Furukawa et al. 1978). It should be noted that tri- and dichlorophenols are also attacked, as well as the fully saturated pentachlorophenol, although at only ca. 10 percent the relative level. The last observation is that 2-chlorophenol is attached, indicating an interesting ortho cleavage mechanism which is being pursued.

Table III. Respiration and substrate specificity of a 2,4,5-T
induced strain.[a]

Substrate	n moles O_2 consumed/min[b]
2,4,5-Trichlorophenoxyacetic acid	69.4
2,4,5-Trichlorophenoxypropionic acid	0.0
2,4-Dichlorophenoxyacetic acid	24.0
2-Methyl-4-chlorophenoxyacetic acid	0.0
Phenoxyacetic acid	1.9
2,4,5-Trichlorophenol	82.5
2,3,5-Trichlorophenol	25.1
2,3,6-Trichlorophenol	2.3
2,3,4,6-Tetrachlorophenol	12.0
Pentachlorophenol	7.1
2,3-Dichlorophenol	11.6
2,4-Dichlorophenol	11.6
2,5-Dichlorophenol	26.6
2,6-Dichlorophenol	8.3
3,4-Dichlorophenol	19.1
3,5-Dichlorophenol	11.3
2-Chlorophenol	16.9
3-Chlorophenol	7.9
4-Chlorophenol	1.5
Succinate	48.0

[a] All concentrations were 1300 ug/ml except succinate (400 ug/
ml) and chlorophenols (13 ug/ml).

[b] Corrected for endogenous respiration rate.

Future Research Opportunities

Future research in biotechnology directed toward environmental pollution will address many areas. Techniques such as
PAMB need to be carefully examined in order to determine the
exact nature of genetic events that occur. This might be
approached by using only a few specific plasmids and genetically
monitoring changes. Alternately one might examine newly
generated strains via restriction analysis and DNA homologies
via specific sequence probes. The evolution and genetics of all
newly created degradative bacterial strains should provide a
fascinating research opportunity, especially when contrasted

with the more highly characterized plasmid groups, e.g., R
plasmids (antibiotic resistance plasmids).

Another important area to be explored is metabolic pathway
analysis of these newly generated strains. It has not been
determined that these strains (PAMB-generated) follow expected
degradative pathways. Additionally, these data could be coupled
with gene localization to yield single gene manipulation.
Lastly, the isolation of the degradative enzymes for halocarbons
may have several useful applications, e.g., broad specificity
dehalogenase for attacking many compounds. Finally, it is
important to continue to build a gene library to permit more
extensive manipulations in the future. This may be easier than
suspected since many of these degradative activities are plasmid
and transposon mediated.

The field of plasmid ecology must be examined to determine
plasmid stability in nature, as well as plasmid mobility, gene
exchange, and recombination, since these interactions will
heavily determine the degree of success of biotechnology as
applied to environmental pollution.

References

1. Chatterjee, D. K., S. T. Kellogg, K. Furukawa, J. J.
   Kilbane, and A. M. Chakrabarty, 1981. Genetic approaches to
   the problems of toxic chemical pollution, In A. G. Walton,
   Ed., Recombinant DNA, Proc. 3rd Cleveland Symposium on
   Macromolecules, Elsevier, Amsterdam, pp. 199-212.

2. Chatterjee, D. K., S. T. Kellogg, D. R. Watkins, and A. M.
   Chakrabarty, 1981. Plasmids in the biodegradation of
   chlorinated aromatic compounds, In S. B. Levy, R. C.
   Clowes, and E. L. Koenig, Eds., Molecular Biology,
   Pathogenicity and Ecology of Bacterial Plasmids, Plenum, New
   York, pp. 519-528.

3. Dorn, E., M. Hellwig, W. Reineke, and H. J. Knackmuss,
   1974. Isolation and characterization of a 3-chlorobenzoate
   degrading pseudomonad, Arch. Microbiol., 99:61-70.

4. Dorn, E., and H. J. Knackmuss, 1978. Chemical structure and
   biodegradability of halogenated aromatic compounds, two
   catechol 1,2-dioxygenases from a 3-chlorobenzoate-grown
   pseudomonad, Biochem. J., 174:73-84.

5. Friello, D. A., J.R. Mylroie, and A. M. Chakrabarty, 1976.
   Use of genetically engineered multi-plasmid microorganisms
   for rapid degradation of fuel hydrocarbons, In J. M.
   Sharpley, Ed., Proceedings of 3rd International
   Biodegradation Symposium, Applied Science, Essex, England,
   pp. 205-214.

6. Furkukawa, K., K. Tonomura, and A. Kamibiyashi, 1978.
    Effect of chlorine substitution on the biodegradability of
    polychlorinated biphenyls, Appl. Environ. Microbiol.,
    35:223-227.

7. Hartmann, J., W. Reineke, and H. J. Knackmuss, 1979.
    Metabolism of 3-chloro-, 4-chloro-, and 3,5-dichlorobenzoate
    by a pseudomonad, Appl. Environ. Microbiol., 37:421-428.

8. Kellogg, S. T., D. K. Chatterjee, and A. M. Chakrabarty,
    1981. Plasmid-assisted molecular breeding: new technique
    for enhanced biodegradation of persistent toxic chemicals,
    Science, 214:1133-1135.

9. Kobayashi, H. and B. E. Rittmann, 1982. Microbial removal
    of hazardous organic compounds, Environ. Sci. Tech.,
    16:170A-183A.

10. Loos, M. A., 1979. Phenoxyalkanoic acids, In Kearney, P. C.
    and D. D. Kaufman, Eds., Herbicides: Chemistry, Degradation
    and Mode of Action, Marcel Dekker, New York, pp. 1-128.

11. McCall, P. J., S. A. Vrona, and S. S. Kelley, 1981. Fate of
    uniformly carbon-14 ring labeled 2,4,5-trichlorophenoxyacetic
    acid and 2,4-dichlorophenoxyacetic acid, J. Agric. Food
    Chem., 29:100-107.

12. Pemberton, J. M., B. Corney, and R. H. Don, 1979. Evolution
    and spread of pesticide degrading ability among soil micro-
    organisms, In Timmis, K. N. and A. Puhler, Eds., Plasmids of
    Medical, Environmental and Commercial Importance, Elsevier,
    Amsterdam, p. 287-299.

13. Que Hee, S. S. and R. G. Sutherland, 1981. The
    Phenoxyalkanoic Herbicides, Vol. 1: Chemistry, Analysis,
    and Environmental Pollution, CRC Press, Boca Raton, FL.

14. Reineke, W., and H. J. Knackmuss, 1978. Chemical structure
    and biodegradability of halogenated aromatic compounds.
    Substituent effects on dehydrogenation of 3,5-cyclohexadiene
    -1,2-diol-1-carboxylic acid, Biochem. Biophys. Acta,
    542:424-429.

15. Tucker, E. S., V. W. Saeger, and S. O. Hicks, 1975.
    Activated sludge primary biodegradation of polychlorinated
    biphenyls, Bull. Environ. Contam. Toxicol., 14:705-713.

Enzymatic Removal of Hazardous Organics from Industrial Aqueous Effluents

Alexander M. Klibanov

Massachusetts Institute of Technology

Abstract

The enzyme horseradish peroxidase can be used to remove toxic phenols and aromatic amines from industrial wastewater. By adding peroxidase and hydrogen peroxide to wastewater, the pollutants are converted to an insoluble form that precipitates out of the water. An important discovery is that easily removed pollutants aid in the removal of more "stubborn" ones from the water. A microbial peroxidase is sought to make the method more commercially attractive.

Research in Wastewater Purification

More than 500,000 tons of industrial aqueous effluents are discharged annually directly into US coastal waters. These aqueous effluents cause concentration of toxic compounds, excessive growth of undesirable organisms and heavy sediments on the nearshore bottom environment, and they poison marine organisms.

The development of novel, efficient approaches to wastewater purification is therefore a problem of great practical significance.

The first area addressed by my research group at MIT is the enzymatic removal of toxic organics from industrial wastewaters.

A second focus of research is a process we have recently developed (Klibanov and Huber 1981) which uses immobilized whole cells containing hydrogenase activity to remove tritium from aqueous effluents of nuclear power plants. It is a useful

process because, first, it detritiates water, and, second, it
produces tritiated water, T_2O, which is used in various
pharmaceutical and biomedical applications. In the future, when
nuclear fusion reactors are developed, a lot of tritiated water
will be needed.

A third area in which we have started recently deals with
the removal of proteins from industrial wastewaters, in
particular from the seafood and meat processing industries.

In the following, I will concentrate on the first area of
our research activity, which deals with the use of the enzyme
horseradish peroxidase for the removal of phenols, aromatic
amines, and other organic chemicals from industrial wastewater.

Enzymatic Removal of Toxic Organics from Wastewater

Phenols and aromatic amines are present in the wastewaters
of a number of industries, including coal conversion, oil
refining, ore mining, dye and organic chemicals manufacturing,
and the production of plastics and resins. In the future, this
problem is likely to become even more severe, because the coal
conversion industry is rapidly growing in the United States,
which posseses more than 50 percent of the world's resources of
coal. Phenol, cresols, and xylenols are the major organic
contaminants of aqueous effluents produced by the coal conversion industry.

Virtually all phenols and aromatic amines are toxic,
particularly to marine life. Phenol is toxic to fish in
concentrations as low as 5 ppm. Moreover, many phenols and
aromatic amines are mutagenic and carcinogenic.

There are certain methods used today to remove phenols and
aromatic amines from industrial wastewaters. These methods
include extraction with organic solvents, microbial degradation,
adsorption on activated carbon, incineration, irradiation with
ultraviolet light, etc. These methods are quite useful, and
certainly practical in certain areas. However, they suffer from
very serious drawbacks such as high cost, incomplete purification, formation of hazardous byproducts, etc. It would,
therefore, be most welcome if a method could be devised that
does not have these shortcomings, and that is what we have
attempted to do.

We have developed a novel method which uses the enzyme
peroxidase to remove phenols, aromatic amines, and other organic
chemicals from industrial wastewater. Before I describe the
method, let me say a few words about the enzyme peroxidase
itself.

Horseradish peroxidase is a remarkable enzyme. It is one of
the most active enzymes known and has a very high specific
activity. It catalyzes the general reaction

$$SH_2 + ROOH \xrightarrow{\text{peroxidase}} S + ROH + H_2O$$

Peroxidase oxidizes different phenols and aromatic amines

$$\begin{array}{c} \text{PHENOL} \\ \text{or} \\ \text{AROMATIC AMINE} \end{array} \quad + \quad H_2O_2 \quad \xrightarrow{\text{peroxidase}} \quad \text{COUPLED PRODUCTS} \quad + \quad H_2O$$

Reaction Mechanism:

$$E + H_2O_2 \xrightarrow{k_I} E_I + H_2O$$

$$E_I + SH \xrightarrow{k_2} E_2 + S^\bullet$$

$$E_2 + SH \xrightarrow{k_3} E + S^\bullet$$

where E - peroxidase, E_I and E_2 - reaction intermediates, S - substrate (phenol or aromatic amine)

Figure 1. Peroxidase oxidizes phenols and aromatic amines with hydrogen peroxide, resulting in coupled products.

with hydrogen peroxide to give certain coupled products (Figure 1). Peroxidase-catalyzed oxidation of a phenol or an aromatic amine molecule produces two free radicals. These free radicals diffuse from the active center of the enzyme to the aqueous solution, where they can react either with each other or with unoxidized molecules of phenols or aromatic amines. This chain reaction can go on for a fairly long period of time, until two free radicals recombine.

The result of this reaction is a series of coupled aromatic products. The major characteristics of these products are low solubilities in water and high molecular weights. For example, oxidation of aniline with hydrogen peroxide and peroxidase yields as the major product the compound Aniline Black (Figure 2). This compound is much less soluble in water than aniline. If one compares the solubility of phenol and dimeric phenol in water -- o, o-diphenol or p, p-diphenol -- the difference in solubility is about three orders of magnitude. In other words, even simple dimerization reduces the solubility by as much as three orders of magnitude.

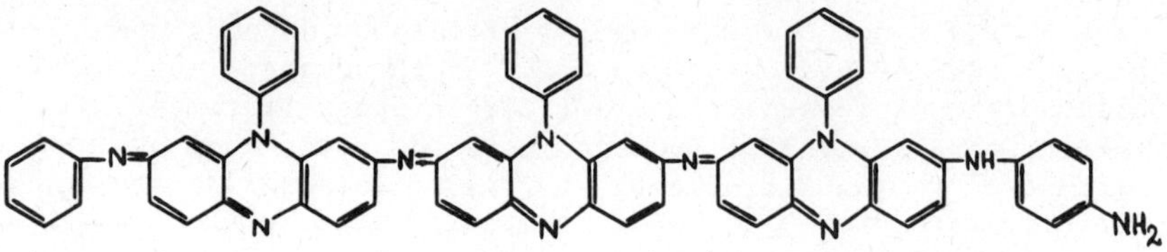

Figure 2. Structure of Aniline Black

This is extremely important for our studies. We use the
enzyme peroxidase to convert water-soluble phenols and aromatic
amines into water-insoluble polyaromatic products. The
rationale behind this approach is that while it is difficult to
remove soluble pollutants from water, it is relatively easy to
remove water-insoluble ones because they will precipitate and
may be filtered or allowed to settle out: in other words, we
use the peroxidase to precipitate pollutants from water.

Testing the Method

To test the theory we attempted the removal of ten known
human carcinogens from water. The experiments were carried out
in relatively simple fashion. We dissolved 100 milligrams/liter
of pollutants in water, which corresponds to 100 ppm, or
approximately the concentration in some industrial aqueous
effluents. We then added hydrogen peroxide and peroxidase to
it. After waiting for a while -- a precipitate is formed in a
few minutes -- we then filtered the precipitate and measured the
concentration of pollutant in the solution. The resultant
concentration of pollutant was compared with the initial
concentration to determine what is called removal efficiency,
i.e. how much has been removed. Obviously, the closer the
removal efficiency is to 100 percent the better the method is.

In virtually all cases, the removal efficiencies were very
high, e.g. 99.9 percent, 99.94 percent, 99.6 percent, 99.9 percent, etc. (Table I). That is, the method seemed to work well
with these pollutants. Next, we wanted to see how the removal
efficiency would depend upon various environmental conditions,
because conditions in real life cannot always be controlled as
well as in a beaker. We singled out one particular carcinogen,
o-dianisidine (Figure 3) and studied the dependence of the
removal efficiency by enzymatic precipitation on different
environmental conditions.

First, we studied how the removal efficiency depended upon
time. As one would expect, the removal efficiency initially
increases with time, then levels off, ending up very close to
100 percent. After one hour, practically all the pollutant had
been removed (Figure 4).

Next, we studied how the removal efficiency depended upon
the concentration of the enzyme. Again, as one would expect, as
the concentration of peroxidase was increased, the removal
efficiency increased to reach the level of nearly 100 percent
(Table II).

Figure 5 shows how the removal efficiency depends upon the
concentration of hydrogen peroxide. If the concentration of
pollutant is expressed in moles/liter in order to compare it
with the concentraton of hydrogen peroxide used, then one can
see that as few as 0.8 molecules of hydrogen peroxide are enough
to precipitate one molecule of the pollutant. Theoretically, it
should be even less than that; it should be 0.5. But, practically speaking, 0.8 molecules is needed to precipitate one

Table I. Removal of carcinogenic aromatic amines from water by horseradish peroxidase and hydrogen peroxide.

Conditions: 100 mg/l aqueous solution of carcinogen,
3 hr treatment at room temperature, pH 5.5,
100 units/l peroxidase, 1 mM H_2O_2 (unless otherwise indicated)

Carcinogenic Aromatic Amine	Removal Efficiency, %
o-Dianisidine (3,3'-dimethoxybenzidine)	99.9 ± 0.1
Benzidine	99.94
3,3'-Diaminobenzidine	99.6
3,3'-Dichlorobenzidine	99.0
o-Tolidine (3,3'-dimethylbenzidine)	99.6
p-Phenylazoaniline	98.5[a]
4-Aminobiphenyl	95.4[b]
α-Naphthylamine	99.7[c]
β-Naphthylamine	98.3[b]
5-Nitro-1-naphthylamine	99.6[d]

[a] pH 4.0, 2 mM H_2O_2

[b] pH 7.0, 5 mM H_2O_2, 1000 units/l peroxidase

[c] 5 mM H_2O_2, 1000 units/l peroxidase

[d] pH 4.5, 5 mM H_2O_2, 200 units/l peroxidase

molecule. This point is important and will be expanded on later.

Figure 6 shows how removal efficiency depends on pH. Obviously, pHs of aqueous effluents vary, so a pH-independent method would be desirable. We deliberately used a concentration of the enzyme which was not sufficient to precipitate the entire amount of the pollutant present so that we could see the differences between different pHs. Fortunately, the removal efficiency is fairly independent of pH ranging from three to nine. It is useful that the enzyme has a broad removal efficiency within this wide pH range.

One of the major problems in pollution control is that some methods (e.g., microbial oxidation or adsorption on activated

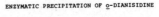

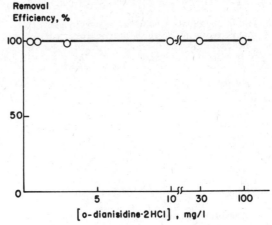

Figure 3. Enzymatic precipitation of o-Dianisidine.

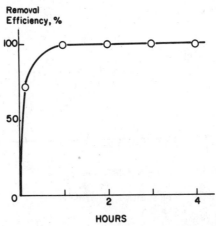

Figure 4. Dependence of the removal efficiency of o-Dianisidine from solution on time.

carbon) work well at low concentrations of pollutants, whereas other methods (e.g., extraction with organic solvents) work well at high concentrations of pollutants. So it was important for us to investigate how the removal efficiency of our enzymatic treatment would depend upon the concentration of the pollutant.

In a relatively broad range of concentrations which vary by a factor of 100, removal efficiency is fairly independent of pollutant concentration. The removal efficiency was approximately the same -- close to 100 percent.

Table II. Dependence of the removal efficiency of o-Dianisidine from solution on the concentration of horseradish peroxidase.

Conditions: 100 mg/l o-dianisidine in 10 mM acetate buffer (pH 5.5), 3 hr treatment at room temperature, 1 mM H_2O_2

Concentration of Horse Radish Peroxidase units/l	Removal Efficiency %
100	99.97
30	99.94
10	93.7
3	86.6
1	82.2
0.3	60.4
0	0

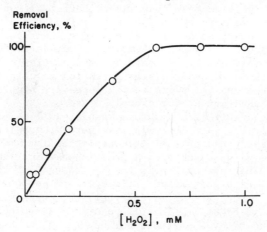

ENZYMATIC PRECIPITATION OF o-DIANISIDINE

Figure 5. Dependence of the removal efficiency of o-Dianisidine from solution on the concentration of hydrogen peroxide.

ENZYMATIC PRECIPITATION OF o-DIANISIDINE

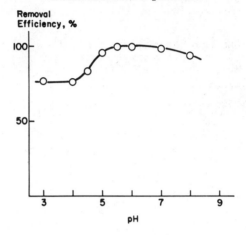

Figure 6.   Dependence of the removal efficiency of
o-Dianisidine from solution on pH.

Removal Efficiencies for Phenols and Anilines

We were then ready to extend our investigations to other
classes of pollutants.   Table III shows the removal efficiencies
for different phenols.   In some cases (i.e., for m-methoxy-
phenol, o-chlorophenol) we had relatively good removal
efficiencies, such as 99.8, 99.6 percent; in other cases the
removal efficiencies were not as good: 53 percent, 85 percent,
84 percent.   The situation is a little better with naphthols
because these are more hydrophobic, therefore the products of
their enzymatic oxidation will also be more hydrophobic and
hence less soluble in water.   Removal efficiencies for all
binuclear phenols are higher than 90 percent (Table IV).

With anilines we had the same situation as with phenols.   In
some cases, the removal efficiencies were very high (98 percent
with m-phenylenediamine, for example), whereas in other cases
they were not nearly as high, for example, aniline -- only 73
percent (Table V).   At that point, we were disappointed that the
method apparently had only limited applicability.

Fortunately, we discovered a phenomenon which removes this
limitation.   That phenomenon is the following: if two pollutants, one with a high removal efficiency and another with a low
removal efficiency, are mixed together, the overall removal
efficiency will be high.   That is, easy-to-remove pollutants
facilitate the removal of those pollutants which cannot be
easily removed.

For example, if o-aminophenol is treated alone, the removal
efficiency is 48.6 percent.   However, if the enzymatic treatment
is carried out in the presence of dimethylphenol, p-phenyl-

Table III. Removal of various mononuclear phenols from water by horseradish peroxidase and hydrogen peroxide.

Conditions: 0.1 g/l aqueous solution of phenol, 3 hr treatment at room temperature, 1 unit/ml peroxidase, 1 mM H_2O_2

PHENOL	OPTIMAL pH	REMOVAL EFFICIENCY[a],%
Phenol	3.5	85.3
Guaiacol (o-methoxyphenol)	5.5	98.0
m-Methoxyphenol	5.5	98.6
p-Methoxyphenol	7.0	89.1
o-Cresol	4.0	86.2
m-Cresol	4.0	95.3
p-Cresol	5.5	85.0
o-Chlorophenol	7.0	99.8
m-Chlorophenol	7.0	66.9
p-Chlorophenol	5.5	98.7
o-Aminophenol	3.5	53.5
m-Aminophenol	5.5	85.3
Resorcinol	3.5	84.1
5-Methylresorcinol	3.5	90.8
2,3-Dimethylphenol	4.0	99.7
2,6-Dimethylphenol	5.5	82.3

[a] at the optimal pH

phenol, naphthol, or naphthalenediol, the removal efficiency of aminophenol increases to 95 percent, 92 percent, 84 percent and 95 percent respectively (Table VI).

The results are even more dramatic with phenol itself. When phenol is treated with the enzyme, the removal efficiency is about 74.6 percent; whereas, if this enzymatic treatment is carried out in the presence of o-diansidine, benzidine or 8-hydroxyquinoline (all of which can be easily removed by the enzyme), the removal efficiencies are all in excess of 99 percent (Table VII).

Why does this happen? What is the mechanism of this phenomenon? The details of the mechanism are not yet clear.

Table IV. Removal of various binuclear phenols from water by horseradish peroxidase and hydrogen peroxide.

Conditions: 0.1 g/l aqueous solution of phenol, 3 hr treatment at room temperature, 1 unit/ml peroxidase, 1 mM H_2O_2

COMPOUND	OPTIMAL pH	REMOVAL EFFICIENCY[a], %
1-Naphthol	4.0	99.6
1,3-Naphthalenediol	4.0	92.1
2-Nitroso-1-napthol	4.0	98.9
2,7-Naphthalenediol	3.5	99.1
p-Phenylphenol	4.0	99.9
5-Indanol	7.0	96.3
8-Hydroxyquinoline	7.0	99.8

[a] at the optimal pH

Table V. Removal of aromatic amines from water by horseradish peroxidase and hydrogen peroxide.

Conditions: 0.1 g/l aqueous solution of aromatic amine, 3 hr treatment at room temperature, 1 unit/ml peroxidase, 2 mM H_2O_2

AROMATIC AMINE	OPTIMAL pH	REMOVAL EFFICIENCY[a], %
Aniline	7.0	72.9
4-Chloroaniline	5.5	62.5
4-Bromoaniline	5.5	84.5
4-Fluoroaniline	7.0	86.4
4-Bromo-2-methylaniline	7.0	84.5
m-Phenylenediamine	7.0	98.6
4,4'-Methylenedianiline	7.0	88.9
N-(1-Naphthyl)-ethylenediamine	7.0	93.3
4'-Amino-2,3'-dimethylazobenzene	5.5	95.5

[a] at the optimal pH

Table VI. Efficiency of the enzymatic removal of o-Aminophenol in the absence and in the presence of other compounds.

Conditions: 0.1 g/l aqueous solutions of o-aminophenol and other compounds, 1 unit/ml peroxidase, 2.5 mM H_2O_2, pH 4.0, 3 hr treatment at room temperature.

POLLUTANT	ADDED COMPOUND	REMOVAL EFFICIENCY, %
o-Aminophenol	None	48.6
o-Aminophenol	2,3-Dimethylphenol	95.1
o-Aminophenol	p-Phenylphenol	92.0
o-Aminophenol	1-Naphthol	84.9
o-Aminophenol	2,7-Naphthalenediol	95.3

Table VII. Efficiency of the enzymatic removal of phenol in the absence and in the presence of other compounds.

Conditions: 0.1 g/l aqueous solutions of phenol and other compounds, 1 unit/ml peroxidase, 2.5 mM H_2O_2, pH 5.5, 3 hr treatment at room temperature

POLLUTANT	ADDED COMPOUND	REMOVAL EFFICIENCY, %
Phenol	None	74.6
Phenol	o-Dianisidine	99.7
Phenol	Benzidine	99.5
Phenol	8-Hydroxyquinoline	99.8

However, some general features are understood. First of all, it should be stated that the free radicals produced from compounds with low removal efficiency apparently react slowly with the unreacted molecules and/or the products are fairly soluble in water. In this case, obviously, when a pollutant which can be easily removed from the system is added, its free radicals can react not only with its own parent molecules, but also with the molecules of other phenols and aromatic amines. The resulting mixed polymers are not soluble in water and readily precipitate.

Other Organic Compounds

It occurred to us that by using this approach we could remove not only phenols and aromatic amines, but also other organic compounds: the free radicals produced from easy-to-remove phenols and aromatic amines would attack other molecules which are not substrates for peroxidase. We have found that this is indeed the case.

For example, if naphthalene, which is a fairly inert molecule, reacts with hydrogen peroxide and peroxidase, nothing will precipitate. In the presence of dimethylphenol, however, as much as 60 percent of the naphthalene can be precipitated.

The results are even more impressive with azobenzene, which is a common chemical intermediate and a suspected carcinogen. Again, it does not react with peroxidase and hydrogen peroxide. However, when treated in the presence of 8-hydroxyquinoline, as much as 99 percent can be precipitated.

The implications of these findings are very important, because real industrial wastewaters never contain just one, but many pollutants. Therefore, even if just a few of these pollutants can be easily precipitated, they hopefully will help to remove the rest. This turns out to be the case, as you will see.

Our work wasn't meant to be just an academic exercise; we wanted to develop a practically useful process. Therefore, the cost of the process shouldn't be prohibitive. How can the cost be reduced? There are two components in our process: hydrogen peroxide and horseradish peroxidase.

As far as hydrogen peroxide is concerned, there is not much one can do; as far as horseradish peroxidase is concerned, however, the story is different. If a crude preparation of the enzyme could be used as opposed to a purified preparation, it would be cheaper. We wanted to establish whether a crude preparation of the enzyme could, in fact, be used. In order to test it out, my technician, Barbara Alberti, who did much of this work and certainly deserves a lot of credit, went to a local supermarket. She bought a pound of horseradish roots which she brought back to the lab and minced in a blender. She then added some water, stirred for a couple of minutes, and pressed it through a cheesecloth. We used the resultant juice as our source for the enzyme. We compared this crude preparation of the enzyme with a purified preparation of horseradish peroxidase obtained from a commercial supplier. We found that the crude enzyme works just as well as the purified product. The removal efficiencies obtained with both preparations are approximately the same (Table VIII).

Experiments on Real Industrial Wastewater

We were then ready to experiment with real industrial wastewater and see how the enzymatic treatment worked. We obtained a sample of industrial aqueous effluents from a chemical plant which produces triarylphosphates, used as flame

Table VIII. Removal of phenols from water by purified and crude peroxidases.

Conditions: 100 mg/l phenols,
1000 units/l peroxidase, 2.5 mM H_2O_2, 21 hr treatment
at room temperature.

Phenol	Removal Efficiency[a], %		pH
	"Sigma" enzyme	Crude enzyme	
Phenol	87.6	89.8	3.5
3-Methylphenol	85.3	99.6	4.0
4-Methylphenol	98.5	95.6	5.5
2-Chlorophenol	99.7	99.7	7.0
2,3-Dimethylphenol	99.6	99.3	4.0
1-Naphthol	97.4	99.6	4.0
2-Nitroso-1-naphthol	96.0	95.0	4.0
8-Hydroxyquinoline	99.9	99.7	7.0

[a] The phenols were assayed by the method of Emerson [17] (except for 4-methylphenol which was assayed in accordance with Arnow [19]) as described in [10]

retardants. In this plant in Ohio there are five ponds with approximately 50 million gallons of these contaminated waste-waters, so the plant gladly sent us a couple of gallons to work with. The material looked exactly as we expected: a yellow, stinking liquid, containing more than 50 different phenols and other chemicals, such as phosphate, triarylphosphates, etc. The treatment the company currently gives to this material consists of two steps. First, bacteria are added to ponds with the contaminated waters. In the course of approximately a week, the bacteria reduce the concentration of phenols by about 80 to 85 percent. The rest of the phenols are then adsorbed on activated carbon. The activated carbon cannot be used right away, since it would overload very quickly, and its regeneration is expensive.

The method is further complicated by the fact that the bacteria will not metabolize the phenols in cold weather; the microorganisms will consume the chemicals only in the summer when it is warm. The company specifically asked us, therefore, to carry out the reaction at a low temperature, e.g. 4°C., because at that temperature the current method does not work at all. The company also asked us to operate our process at pH 7, since if the pH is lower the solution will corrode pipes and other metal equipment.

Accordingly, we took a liter of the sample, brought the pH up to 7, put it in the refrigerator, and added the enzyme and H_2O_2 to the cold solution. The sample was left in the refrigerator for one day. After that, we analyzed the concentration of phenols remaining. We found that after just one day at $4^{\circ}C$, we obtained a 96 percent removal efficiency, as compared to their 80 to 85 percent after seven days and only at a much higher temperature.

We had thus proved the technical feasibility of our process. We then wanted to determine if the method could in fact be economical.

Cost Effectiveness

We compared the cost of the materials in our process with that of an existing process, the FMC process, for the oxidation of phenols. The FMC method does not oxidize most substituted phenols or aromatic amines, but degrades phenol itself. The method operates by taking a solution of phenol, adding hydrogen peroxide to it, and using an iron catalyst to oxidize the phenol to carbon dioxide and water. The problem with the method is that since phenol is oxidized to carbon dioxide and water, the process requires a lot of hydrogen peroxide. In fact, it requires 14 molecules of hydrogen peroxide per molecule of phenol. We ignored the cost of the iron used and estimated only the cost of hydrogen peroxide. It turns out that using the FMC method to treat 1,000 liters of wastewater containing 100 ppm of phenols would cost about 85 cents.

With our method there are two components. The first is hydrogen peroxide. It is significant that we use very small amounts of hydrogen peroxide, because we don't oxidize the pollutant to CO_2 and water; we just precipitate it. That is an important consideration. The second component is horseradish peroxidase. On the basis of the current wholesale price of horseradish, the cost of our process to treat 1,000 liters would be approximately 70 cents, compared to FMC's 85 cents. Although our process is thus not much cheaper, there are two important differences between our method and FMC's method:

● First of all, the FMC process is only effective with phenol itself; it cannot treat most substituted phenols and aromatic amines and other organic chemicals. Our process can.

● Second, the FMC process oxidizes phenol to carbon dioxide and water: hence, phenol disappears after it is oxidized. We do not do that; we precipitate the phenols and aromatic amines, and as a result produce a lignin-type polymer which can be burned to produce heat or electricity. This precipitate has significant fuel value. For example, we have calculated that if you take an average-sized commercial coal gasification plant and estimate that you can precipitate 80 percent of the phenols it produces, you will obtain about 100 tons of this precipitate per day, which has a substantial fuel value.

A Microbial Peroxidase

So it seems that our method can have certain advantages. However, I would like to point out here that despite the fact that from an economic standpoint we approximately break even, it should be stressed that the enzyme is approximately 80 percent of the cost of materials in this process. So, first of all, it will be helpful to find a microbial peroxidase which can perform as well as the horseradish enzyme, but is cheaper. Secondly, the supply of horseradish in this country is rather limited.

There are indications that we can find a suitable microbial peroxidase and we are working in this direction now. If we are unsuccessful with this approach, then horseradish peroxidase can be cloned and this gene expressed in Escherichia coli or a similar microorganism.

In closing, I would like to say that at this point, it is hard to say whether our method will ever be in practical use, but there are indications that it might be.

References

1. Arnow, 1937. J. Biol. Chem., 118:531.

2. Emerson, 1943. J. Org. Chem., 8:417.

3. Klibanov, A. and Huber, 1981. Biotechnol. Bioeng., 23:1537.

Microbial Problems, Solutions, and Trends in Industrial Waste Treatment

Robert L. Wetegrove

Nalco Chemical Company
Oak Brook, Illinois

Abstract

Disturbances in the biological balance of an industrial
wastewater treatment system can be manifested in several ways,
including (1) destruction of floc; (2) bulking sludge, which
will not compact because filamentous bacteria have proliferated
and extended into the surrounding fluid; and (3) floating
sludge, formed when gas attaches to floc and floats it to the
surface. Solutions for these problems exist, but the problems
are best handled with an understanding of the interrelatedness
of the operations of the unit, microbiological processes, and
polyelectrolyte applications. Future trends in wastewater
treatment include emphasis on making existing plants more
efficient rather than on building new plants; dewatering sludge
with a twin belt press; and development of energy-efficient
processes.

Efficient removal of dissolved and suspended waste matter
from industrial effluents requires a coordinated understanding
of unit operations, microbiological processes, and polyelectrolyte applications.

All industrial wastewaters carry unique and complex mixtures
of dissolved, colloidal, and suspended matter. The nature of
the waste depends on the uses for which the water in the
factory, refinery, mill or plant was employed. The water may
contain insoluble inorganic or organic solids, oils, and dissolved inorganics and organics. Any of these components may be

275

Figure 1.   Primary clarifier.

toxic to or readily assimilable by living organisms. In the
first stage of effluent purification (Figure 1), known as primary treatment, settleable or floatable solids are concentrated
and removed by physical processes such as gravity clarification
or dissolved air flotation. These processes are commonly
enhanced by the use of organic polyelectrolytes which aggregate
suspended particulates.

Removal of the remaining soluble organic and inorganic waste
products is done in secondary, or biological, treatment. In
this operation the goal is to create and maintain conditions in
which the proper microbes will grow on the soluble organics and
reproduce to form settleable or adhesive masses. Figures 2, 3,
and 4 show an aerated lagoon, submerged aeration activated
sludge, and trickling filter systems.

The biocommunity in a well-operated secondary waste treatment system comprises an extremely complex population of various
physiological and morphological types. Each member of the
population fills a biochemical niche in a sludge floc micro-
environment. The cells in a floc are held together by extracellular capsular polysaccharides produced by microorganisms in
the floc. The polysaccharides may also serve to adsorb and
concentrate charged colloidal molecules from the surrounding
fluid (Figure 5). Protozoa and metazoa will often become
established members of the populations, anchoring themselves to
the floc as filter-feeders or browsers (Figure 6). Development
of this state in a floc is the goal of conventional aerobic

Figure 2.   Aerated lagoon.

Figure 3.   Submerged aeration activated sludge.

processes because it provides efficient removal of soluble organics and leaves a clear effluent from the secondary clarifier (Figure 7). Thickening, dewatering, and disposal of the biological solids is also most efficient when the secondary treatment system is healthy.

Figure 4. Trickling filter.

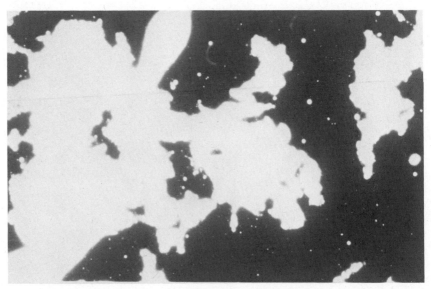

Figure 5. Activated sludge, 125X magnification.

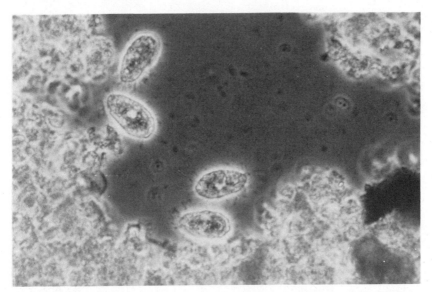

Figure 6.   Protozoa in activated sludge, 450X magnification.

Figure 7.   Secondary clarifier effluent.

Upsetting the Biological Balance

Upsets in the biological system usually lead to an increase in suspended solids and soluble organics in the effluent. An "upset" can be caused by any change in the type or concentration of organic or inorganic waste, a sudden change in pH, temperature, oxygen concentration, trace nutrient level or any other factor affecting the population balance. The cause or causes of floc destruction can be as obvious as a toxicant spill, or may be subtle and quite baffling, but the result is dispersed (deflocculated) growth (Figure 8). Under these conditions single cells or small clumps of cells develop a substantial net negative surface charge which must be neutralized by a cationic coagulant before the bacteria will aggregate into floc large enough to settle easily.

"Bulking" sludge is sludge which will not compact because filamentous bacteria have proliferated and extended beyond the margins of the sludge floc into the surrounding fluid like stiff threads (Figure 9). The poorly compacted floc soon fills the clarifier and spills into the effluent. A commonly accepted explanation for filamentous bulking suggests that it is promoted by a nutrient imbalance. In this sense, nutrients included degradable organics, nitrogen, phosphorus, trace metals, and oxygen. The relative lack of any of these nutrients will favor growth of filamentous bacteria because of the bacteria's greater

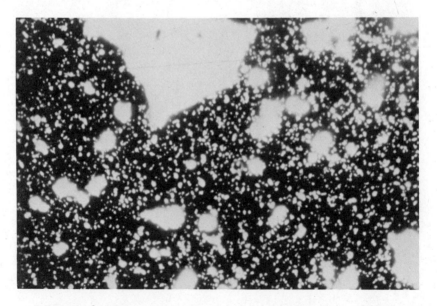

Figure 8. Dispersed floc in activated sludge, 125X magnification.

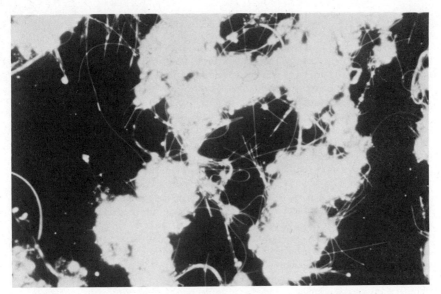

Figure 9.   Filamentous bulking sludge, 125X magnification.

surface-to-volume ratio and because the filaments extend from the floc out into the surrounding fluid. This gives the filaments a competitive advantage in obtaining the limiting nutrient.

There are basically two approaches to control filamentous bulking. Neither is always successful. The first approach, consisting in establishing a nutrient balance, is time consuming. The nutrient (or nutrients) out of balance may not be easy to identify because at least three sludge half-lives are normally required for cell turnover sufficient for a population shift to become evident. Insufficient oxygen for a plant's organic loading is a common cause of bulking.

The second approach to control of filamentous bulking is the application of oxidizing toxicants at low levels. The idea is to inactivate the filaments selectively in short-time contacts with the toxicants; this is possible due to the filaments' large surface area and exposed position.

Floating sludge can occur if biologically formed gas in the bottom of a clarifier attaches to floc and floats it to the surface. The gas may be hydrogen, carbon dioxide, hydrogen sulfide, methane, nitrogen, or combinations thereof, depending upon conditions. In all cases the occurrence of this phenomenon is caused by an anaerobic environment in the clarifier. The best solution is to eliminate the anaerobic conditions by more efficient sludge removal or mechanical alterations to the rakes or pumps.

The Future of Industrial Wastewater Treatment

These problems with biological waste treatment systems are obviously complex and interrelated. They are best solved with the help of a water treatment expert who understands the unit operations, microbiological processes, and the physical/chemical role of polyelectrolyte applications. Trends in industrial wastewater treatment may be divided into three related areas.

First, construction of waste treatment plants will be minimal in the forseeable future, and emphasis will be on efficient operation of existing facilities. As experience is gained in waste treatment, unit and plant managers will become increasingly aware of the relationships between actions in one part of the system and chemical and operational costs in another. For example, a spill of toxic material in manufacturing may kill bacteria and cause sludge deflocculation in the secondary treatment system. (Compare Figures 10 and 11). This in turn will result in poor removal of soluble waste and increased chemical costs for coagulants and flocculants for clarification and sludge dewatering. Suppliers of chemical waste treatment plants who understand these relationships and who are able to apply a systems approach will provide more value to their customers than suppliers who focus just on the per pound cost of the chemical in a single waste treatment unit operation.

A second trend is dewatering sludge with a twinbelt press. This has become the method of choice for the dewatering step in sludge disposal (Figures 12 and 13). A well-designed twinbelt

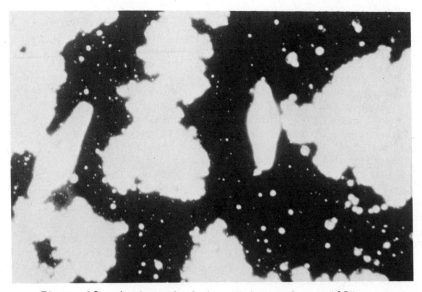

Figure 10. Activated sludge with rotifers, 125X magnification.

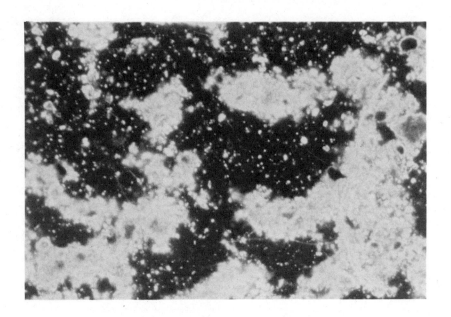

Figure 11. Dispersed Floc in Activated Sludge, 125X magnification.

Figure 12. Waste activated sludge.

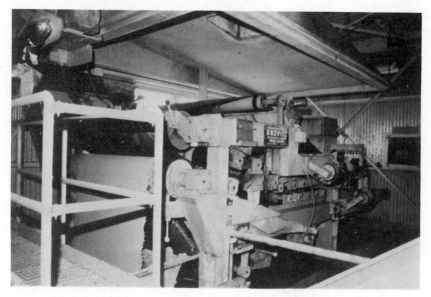

Figure 13.   Sludge dewatering machine.

press is small and inexpensive compared to other mechanical
devices in a waste treatment plant. Maintenance and energy
requirements are also low. However, the sludge must be conditioned with organic polyelectrolytes before it will release
water and withstand the high pressure and shear of the press.
Use of the twinbelt press is expanding rapidly, and product
development is in its infancy, but it is already becoming
apparent that there are definite connections between operations
of biological systems and the relative charge and molecular
weight of the polymer products best suited for dewatering by the
twinbelt press. This again points to the need for a systems
approach to polyelectrolyte supply.

     A third and somewhat longer-term trend in wastewater treat-
ment is the development of processes that are energy efficient
or which yield an energy return. Methanogenic anaerobic digestion of municipal sewage sludge has been practiced for years and
can yield up to about 40 percent of the energy needs of a waste
treatment plant. The application of anaerobic treatment technology to industrial wastes has been slower, however, because
industrial waste treatment is newer, the waste flow and composition are variable and likely to be toxic to the more sensitive
anaerobes, and the processes are more complex to operate. However, as energy has become more valuable and waste treatment
operations more complex, industrial anaerobic waste treatment
schemes are beginning to appear. One of these is claimed to
generate enough methane value to pay for its own operation and
to provide a return on investment for its construction. The

complex and sensitive nature of this and other anaerobic biological processes will create a demand for very efficient chemically-aided primary treatment to remove potentially toxic solids and flotable substances.

In summary, trends in the biological treatment of industrial wastewater are toward more complex and efficient treatment systems. These changes will favor water treatment chemical suppliers represented by technically competent field sales forces. In addition, as futher studies on the basic biology of the anaerobic digestion process are conducted, more efficient systems may result.

Treatment Technologies for the Removal of Biochemical Oxygen Demand

Bruce M. Poole

Sea Plantations Environmental Services, Ltd.
Salem, Massachusetts

Abstract

The removal of biochemical oxygen demanding compounds (BOD) from municipal and industrial wastewaters has become an important goal. Pollution by BOD reduces the waters' capacity to carry oxygen, increases the size of certain bacterial populations, and generally degrades the environment. Treatment systems have been developed that utilize physical-chemical techniques such as sedimentation, filtration, and biological oxidation. The relative efficiencies and operational theories underlying these systems are reviewed in this paper. Specifically, biological oxidation in activated sludge, filters, and ponds are discussed to illustrate conventional techniques and to extrapolate the future potential of waste treatment by microorganisms.

The removal of biochemical oxygen demanding compounds - referred to below simply as BOD - from wastewater has been mandated by regulatory agencies for over ten years to protect the environment, since such compounds can degrade streams, rivers, and oceans if allowed to exceed certain concentrations. Waste product-induced chemical changes result in lowering the water bodies' biological productivity and restricting biological diversity. Areas overloaded with waste products often become turbid, low in dissolved oxygen, and eutrophic because of the initial development of bacterial blooms and the resulting

release of nutrients. To prevent pollution and restore the
health of many rivers, the National Permit Discharge Elimination
System (NPDES) was enacted. Industries and municipalities were
required to treat their wastes to a level that could be assimi-
lated without stress to the ecosystem. Naturally, the limits
set for large rivers and the oceans were influenced by the
dilution factor, resulting in many cases in the need for little
treatment.

All municipal wastewaters are characterized by high BODs due
to the presence of solids and soluble fractions of digested
food. The wastes typically contain carbohydrates and proteins,
with minor amounts of fats, fiber, and nitrogenous residues.
Since all biodegradable substances are metabolized by bacteria,
they exert an oxygen demand. Pure domestic waste has a five day
BOD of 200 mg/l (Fair, Geyer, and Okun 1968). Industrial wastes
from paper mills, tanneries, chemical plants, and rendering have
BODs that range from 500 to 5000 mg/l (Nemerow et al. 1978).
The treatment techniques for BOD depend upon whether the BOD is
solid or soluble and the degree of its biodegradability.
Cellulose, for instance, takes much longer to decompose than
simple polysaccharides such as maltose. Biological oxidation
can also be inhibited or interrupted by toxins, heavy metals,
lack of oxygen or high pH. Since many treatment systems rely on
microorganisms to reduce BOD, conditions in the wastewater
stream must be altered to accommodate the growth and reproduction of the microorganisms. Many municipal plants have and
require industrial pretreatment programs that reduce the heavy
metal ion concentrations in incoming wastes and require the pH
of incoming waste to be adjusted. Both of these procedures
prevent the incoming waste from poisoning the microorganisms in
the digestor.

Treatment

Treatment of wastewaters with high BODs includes a variety
of physical, chemical, and biological techniques. The simplest
unit operation for the removal of particulate matter is direct
sedimentation. Since any solid heavier than water will eventually sink in the medium, clarifiers have been designed to
remove sludges from wastewater. Most municipal installations
include primary clarification and demonstrate removal
efficiencies of 40 to 50 percent. With more potent industrial
wastewaters, chemical coagulation and flocculation are used to
remove a greater percentage (60 to 80) of solids and soluble
BOD. The primary coagulants used are the trivalent metal salts,
such as alum and ferric chloride (Eye and Liu 1971). At
appropriately high pHs these electrolytes form polynuclear
colloidal metallic hydroxides. The complexes are visible to the
naked eye as floc and settle rapidly. Since the floc is a
gelatinous charged particle (-), many fine solids are entrained
in the matrix. Positively charged molecules can also be com-
plexed by sedimentation in a clarifier. These systems operate

with relatively low detention times, usually two to four hours. The use of polymers decreases detention time and increases efficiency.

Rapid sand filtration can be used after clarification to screen out fine particles and colloidal substances that do not settle. Fine silica sand or graded materials of different sizes and densities capture and retain these substances. The filters are cleaned by backwashing the media and returning the solids to the clarifier (Culp and Culp 1974). In tannery wastes, where a large proportion of the impurities is colloidal protein from collagen and hair, rapid sand filtration removes an additional 10 to 20 percent. Four hundred to six hundred mg/l of soluble BOD material can remain in the wastewater and therefore still require biological treatment before they can be discharged into the waterways (Poole 1982).

The conversion of soluble BOD into carbon dioxide and bacterial cell mass, which can then be settled out of solution, represents the standard approach to the reduction of low-level BOD. Municipal treatment facilities operate activated sludge or trickling filters as secondary treatment. In activated sludge, dense suspensions of bacterial solids, protozoans, rotifers, and fungi are aerated and fed clarified wastewater. The rate of BOD uptake is dependent upon the oxygen transfer rate, temperature, cell biomass, and substrate. Effluent from activated sludge proceeds through a secondary clarifier for the removal of bacterial cells and debris. The time required for stabilization of a large percentage of the organic matter is related to the type of organics present. Normal domestic wastes with excess fat, carbohydrates, and protein can be broken down in six to eight hours.

In aerobic decomposition, substrate changes can cause a succession in the genera of bacteria, particularly in multiple series units or extended aeration units. The initial populations of Escherichia coli, Pseudomonas, and Zoogloea ramigera utilize carbohydrates and fats readily. The remaining protein-aceous wastes support the growth of Alcaligenes, Flavobacterium, and Bacillus. As bacterial cell masses increase and begin the die-off phase, fungi, ciliates, and rotifer populations increase. Food/microorganism ratios are maintained through the recycling of the secondary sludge, which is rich in micro-organisms. Industrial wastes containing polysaccharides and aromatic hydrocarbons develop specialized bacterial populations with extracellular enzymes which split large molecules into absorbable fractions. Properly designed systems achieve 80 to 95 percent operational efficiencies and produce an effluent with 10 to 30 mg/l BOD (McKinney 1962).

Trickling filters and rotating biological contactors are fixed bed reactors where bacteria grow on surfaces such as stone, plastic, or fiberglass and metabolize components in the waste stream as it passes by. This type of reactor is more stable than activated sludge but requires more area and strict

hydraulic loading limits (Eckenfelder 1966). Air must
constantly be introduced to the bacteria through trickling,
rotation of media, or intermittent draining. Because the
bacteria established on surfaces are stable and can withstand
shock loads or food scarcity better than suspended bacteria,
these systems are used in the pharmaceutical and chemical
industries, where complex wastes require specific bacteria.
Trickling filters also produce less settleable solids and
accomplish a greater percentage of complete oxidation.

   Oxidation ponds have been used for the longest period of
time to treat BOD. Many different reactions occur in this type
of system concurrently, and the mechanisms are not completely
understood. Aeration is not provided, but low oxygen levels are
maintained by algae in the lagoon. A balance exists between the
amount of oxygen that allows bacteria to metabolize wastes and
release nutrients and the algae that feed on these nutrients.
Cyclic fluctuations of oxygen, wastes, and nutrients occur each
day. Typically, the lagoons accumulate sediments and reach a
point where the anaerobic decomposition in the sediments
exhausts the water's capacity to replenish oxygen. At that
time, treatment is inefficient and odors are produced. When
used as final or polishing ponds, the systems can reduce BOD by
90 to 98 percent and yield an effluent that is "conditioned" and
already supporting algae. Natural populations of zooplankton,
such as Daphnia or copepods, can establish themselves in these
systems (McKinney 1962).

   Table I. Removal Efficiencies of Various Treatment
Operations (Fair, Geyer, and Okun 1968)

Process	BOD (percent)	Suspended Solids (percent)
Simple sedimentation	40–50	40–60
Chemical precipitation and sedimentation	60–80	60–70
Filtration (after above)	70–85	80–95
Activated sludge (after sedimentation)	80–95	80–95
Trickling filters (after sedimentation)	75–95	75–90
Oxidation ponds	90–95	85–95

One such oxidation pond used for the treatment of tannery waste BOD developed dense blooms of a pink blue-green algae, Rhodospirillium. Since blue-green algae can absorb organics directly, it was not surprising to find this alga directly utilizing long chain fatty acids and nitrogenous organics. The alga, being more specific in its uptake, was more effective than bacteria in treating the waste stream, reducing total Kjeldahl nitrogen, oil, and grease, as well as BOD.

The Role for Biotechnology

At this time, BOD treatment technology still depends upon natural populations of microorganisms to utilize the substrates offered as wastes. Reliance on mixed populations and natural seeding results in many problems during daily operation. A crashed reactor population can take weeks to recover and resume adequate treatment. Other organisms such as slime molds, branching fungi and predators may interfere with the treatment by clogging filters and competing with desirable microbes. Biological treatment of wastewaters is still a very inexact science, especially in terms of microbiology. Recent work on the development of specific microbes tailored to provide extra enzymes and different properties indicates the potential of genetic engineering for wastewater treatment. Bacteria custom-fitted for certain waste substrates are more efficient for mixed populations that undergo constant successional flux. The design and use of such microorganisms will greatly improve the technology for the treatment and reduction of biochemical oxygen demanding agents.

References

1. Culp, G.L. and R.L. Culp, 1974. New Concepts in Water Purification, Van Nostrand Reinhold Co., New York, NY.

2. Eckenfelder, W.W., Jr., 1966. Industrial Water Pollution Control, McGraw-Hill, New York, NY.

3. Eye, J.D. and L. Liu, 1971. Treatment of wastes from a sole leather tannery, Water Pollution Control Federation J., 43:2291-2302.

4. Fair, G.M., J.C. Geyer and D.A. Okun, 1968. Water and Wastewater Engineering, Vol. 2, John Wiley and Sons, Inc., New York, NY.

5. McKinney, R.E., 1962. Microbiology for Sanitary Engineers, McGraw-Hill, New York.

6. Nemerow, N.L. et al., 1978. Proc. 33rd Industrial Waste
    Conference, Purdue University, Lafayette, IN.

7. Poole, B.M., 1982. Enhanced Clarification of Tannery
    Wastewater, Journal of the American Leather Chemists Assn.,
    77:24-34.

Author Index